Solar Home Design Manual for Cool Climates

Shawna Henderson and Don Roscoe

from Routledge

First published 2010 by Earthscan

First published 2017 by Routledge
2 Park Square, Milton Park, Abingdon, Oxon OX14 4 RN
52 Vanderbilt Avenue, New York, NY 10017

Routledge is an imprint of the Taylor & Francis Group, an informa business

ISBN 978-1-84407-969-8 (pbk)

Text pages designed and typeset by Ideas Ink Design
Cover design by Yvonne Booth

A catalogue record for this book is available from the British Library

Library of Congress Cataloging-in-Publication Data has been applied for

CONTENTS

ACKNOWLEDGEMENTS

Putting together a manual like this involves the help and cooperation of many people and organizations. What you are holding in your hands is a revision of a revision. Back in 1986, members of Solar Nova Scotia created a manual to go with a course on solar greenhouse design and construction. In 1992, the manual was revised and updated to better suit the course participants were really interested in, namely passive solar home design. This is now the new global edition expanded to cover regions of the world other than Canada. It has been updated with new information about materials, assemblies and building techniques. There is information about solar thermal and photovoltaics, as well as how to make your home 'solar ready'. At the end there are a series of international case studies that showcase different ways of using solar energy in the home.

This manual is designed to provide information about the design and construction of solar homes. Every effort has been made to make this manual as complete and accurate as possible within current codes of practice, but no warranty or fitness is implied. The information is provided on an 'as is' basis. If expert assistance is required, the services of a competent professional should be sought.

Shawna Henderson and Don Roscoe

INTRODUCTION

If you want an inexpensive, environmentally sound source of energy for your house, you need look no further than the sun. Solar heat is not subject to rate increases, is totally renewable, pollution-free and requires no fancy technology. It is here for you today, and can easily provide up to 50 per cent of your space and water heating requirements.

Conservation of non-renewable resources and cost savings are the prime reasons people become interested in solar energy. However, there are others that are no less important – the most obvious being the impact that our energy use has on the global environment. You can minimize this impact by consuming as little as possible of the cleanest available energy. Other important considerations are the amount of energy used to produce your home's building materials, the environmental impact of their manufacture and eventual decay. Buying locally produced materials not only boosts the regional economy – it also reduces the environmental cost of product transportation. Other lifestyle choices, such as recycling, composting, water conservation, reduced automobile use and the elimination of toxic products in and around the home all reduce the ecological footprint of your household. A house is the biggest single investment most of us will make in our lives – the way it is built and how it operates can reflect a long-term investment in both the building and the planet.

The initial cost of building an energy efficient house or renovation may be slightly higher. However, this cost is offset over the long term by the payback in low fuel bills.

Using the sun for home heating is not a new idea. Twenty-five hundred years ago, Socrates designed houses to take advantage of the sun for heating and cooling. Many traditional types of architecture collect and store the sun's heat for use at night. For some people, the phrase 'solar power' evokes images of experimental buildings complete with 45-gallon drums filled with water and an aging hippy or two. Although 45-gallon drums filled with water are an effective means of storing heat collected from the sun, there are other methods that are not so obtrusive. Solar houses don't have to look like spaceships, they can work with any style. This book introduces you to the practical potential of solar energy.

There are several strategies we can use to maintain a comfortable thermal environment within a solar home. The first priority is **energy conservation**. When a home's heating requirement is reduced to its lowest practical level, it becomes feasible to meet a large portion of that demand with solar techniques. It is generally easier and cheaper to control heat losses than to supply solar gain. In the same vein, reducing electrical loads – even if you are not generating your own power – is a high priority, as it is easier and cheaper to save a watt than it is to make a watt.

Reducing heat loss in the winter and heat gain in the summer is the best way to minimize your heating and cooling costs. Heat loss cannot be eliminated, but it can be significantly reduced. An energy efficient house reduces the amount of heat required to maintain a constant temperature. This is achieved through the use of high levels of insulation, airtight construction techniques, controlled ventilation and an efficient heating system. A solar house will not function effectively unless it is designed to contain the collected heat – otherwise, the purpose of capturing the sun's energy as heat is defeated. An energy efficient house is not necessarily a solar house, but a solar house must be energy efficient.

For the purposes of this book, we will ignore the term 'hybrid' when talking about passive solar houses fitted with fans or blower units. This is because 'pure' passive systems don't work well in houses designed with interior walls (which block natural convective air movement), and they also require a certain level of hands-on monitoring to perform adequately. By installing a mechanical means of moving the heated air, you can supply that heat to areas where it is needed, and you can control the climate in your house automatically. This means the house is much more efficient and comfortable.

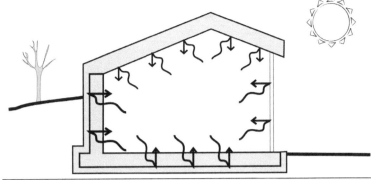

WHY A SOLAR HOUSE MUST BE ENERGY EFFICIENT:
Heat is lost through all exterior surfaces – the foundation, walls, roof, windows and doors.

Installing high performance windows might cost more than using ordinary double-paned glass, but those same windows could save significantly in annual heating costs, and provide higher comfort levels by eliminating cold spots. A small investment in insulation can add to savings on heating costs as well. Heat loss from leaky door frames, windows and under-insulated areas of your house will, in the long run, cost you more, than the initial investment in energy efficient construction. The extra money spent on improving the building envelope will result in lower heating requirements, so you can use a smaller, less expensive heating system. In many cases, the overall construction budget stays the same, where the money is spent varies.

A cool climate can also be described as a 'heating' climate, meaning there are space heating requirements for most of the year. In most cool climate regions, an energy efficient, well-insulated home with properly oriented windows does not need air conditioning during the summer. A few well-placed opening windows, overhangs or shading devices on south and west facing windows can eliminate many overheating problems. These measures are far less expensive, much more energy efficient and environmentally sound, than investing in a mechanical air conditioning unit.

solar basics

To be effective, a solar house must collect, store and distribute heat from the sun. So, after energy conservation, the next priority for your solar house is the supply side. How can you maximize the amount of sunlight that can be converted to heat, and what will you do with that heat once you have it?

There are three 'streams' of solar energy technologies: passive solar design, solar thermal heating and photovoltaics. Passive solar design relies on sunlight coming through windows for space heating and can easily be incorporated into a house design without much extra cost. A solar thermal system uses external solar collectors to heat fluid or air for space and water heating. Although these systems can increase your capital cost, they have relatively short payback periods. Photovoltaic systems use special semi-conductor materials to change light into electricity, and are currently very expensive with long paybacks in most cases. Both solar thermal systems and photovoltaics systems can be incorporated into the house design, or added on after the house is built.

A passive solar house acts as one big heat collector and storage unit. A truly 'passive' house relies on manually operated vents and natural convective currents to move heat. Remember the saying 'hot air rises'? Well, it does. And cold air falls. That's a convective current in action. When you rely on natural convective currents to move heat around, they don't always move the heat fast enough to maintain a comfortable, even temperature. The result: the air in the bottom half of the enclosed space – where you live – remains cool, while the proverbial bats in your belfry swelter in the heat. Temperature stratification can be solved by 'forcing' convective air currents; fans or blower units move the air at a faster rate to even out the temperature in the space.

TO BE EFFECTIVE, PASSIVE SOLAR STRUCTURES MUST DO THREE THINGS:

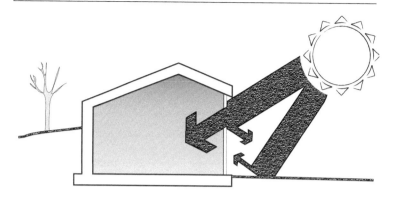

1 Collect the sun's energy in winter, not in summer

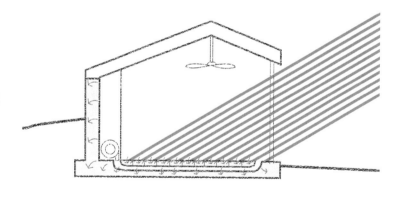

2 Store the sun's energy

It's much easier to plan and design a house to capture the sun's energy for winter heating needs than it is to avoid it in the summer. Without including summer sun angles in your design, you can easily create overheating problems.

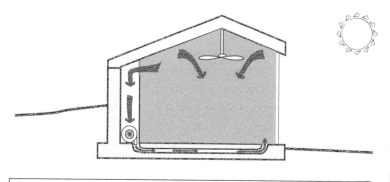

3 Distribute the stored energy

Although we have used the term 'house', which implies 'single family residence', the concepts we have laid out in this manual are applicable to any type of building. The physics stay the same even when the scale and use change – of course, the bigger the project the more complex it becomes.

Active solar systems use machines to move heated fluid or air. They require collectors, storage media, pumps, blowers and control systems. The cost and complexity of active systems suggest that they be considered after conservation and passive techniques have been used to the fullest. On the other hand, a properly sized solar thermal system allows more precise control of internal temperatures, with somewhat lesser demands placed on the design of the house. Solar thermal systems can be the most effective way to use the energy from the sun if your house (or building site) doesn't have the proper orientation or physical setting for passive solar design.

Solar energy is well-suited to providing domestic hot water. Solar domestic hot water (SDHW) systems are reasonably priced, easily installed and can account for over half of your annual hot water needs. They are best suited to a household with high hot water consumption. SDHW systems are also practical when used to heat pools and hot tubs.

Electricity can be made directly from the sun with photovoltaic (PV) systems. PV costs are dropping rapidly but are still relatively high. On the other hand a PV system may be cost effective in meeting your electrical needs if your home or cottage is a considerable distance from the power grid.

ABOUT THIS BOOK

We saw a need for a book that simply and clearly explains the principles of using solar energy so that anyone building a new house, or renovating an old one, can incorporate one, or several, aspects into their design.

This is a process-oriented manual, taking you through the process of designing a solar house from the ground up. It is not a technical manual. There are very few numbers and very few calculations. Several excellent technical books and online resources are available for those interested in number-crunching. This manual is also a basic course in conservation and sustainable house design. Current housing practices have a major impact on the environment. We can either go on using huge amounts of non-renewable resources, with their attendant financial and environmental costs, or we can find different ways of meeting our housing needs.

In California in 2007, one neighbour was informed that they would have to cut down a stand of venerable redwoods as the trees were blocking the other neighbour's 'right to light', that is solar access for a photovoltaic system. This is a classic case of poor planning, as the neighbour with the shading problems certainly knew the trees were there before installing the photovoltaic system!

Photo courtesy of Greg Allen

starting from the ground up

This book is focused on the process of designing a solar house from the ground up. The first aspect of the design process is to take stock of what you have. In this case, it is the bit of land upon which you will build, or the house you are renovating and its site.

A solar house works with its site to maximize the gain of available free energy. This means learning about your site at many levels: from the 'big picture' to the microclimate. This includes observation, research and interpretation of your findings. How do the prevailing winter winds affect your site? How can you take advantage of the existing features of your site to modify the effects of winter winds while magnifying the amount of solar gain? How can you work with your rooflines to get the most out of a solar thermal or photovoltaic system? Are there trees on the neighbouring property or an empty building lot to the south that could reduce your solar access in the future? The performance of a solar house is dependent on how well you know your site and how adeptly you use it. Overleaf is a summary of some of the factors influencing performance, plus a table comparing cities around the world.

Once you understand your site's advantages and limitations, you have a context in which to design your house. This book outlines the basic principles of solar house design, as well as ways to explore and determine what you want to design into (and out of) your new house. The section on house design focuses on the basics of incorporating solar into a house – including how to plan for solar thermal and photovoltaic systems. Also included are ways to take advantage of natural daylight and the relationship of the house with the site.

It may be important to remember that the design process is not a straight-line 'A-to-B' process. There are many aspects to consider, and many of those aspects are interconnected in subtle ways. Lifestyle, aesthetics, physics, natural cycles, environmental concerns, health, wealth, social values, family values, personal values, thought processes, hierarchies of meaning, patterns and geometry … the list can go on and on. When you design a solar house, you are adding another set of considerations.

There are many excellent books and online resources dedicated to house design and layout.

MATERIAL WORLD

When making choices about products and assemblies you may wish to consider these aspects:

- *Embodied energy and embodied pollution: how much energy is used and how much pollution is created by the material or assembly during the following stages?: mining or harvesting manufacturing, transport, installation, demolition and disposal/recovery*

- *Sustainably harvested wood or plant material: look for local sources start within 100 miles of your site and move out from there*

- *High post-consumer recycled material content*

- *Material/product service life*

- *Potential for future recycling of material when service life completed*

- *Assembly service life*

One aspect of energy efficiency measures in new and retrofit projects that is not well-addressed is the value of 'non-energy benefits' (NEBs). Energy saving programmes generally ignore these, mainly because they are difficult to quantify, yet surveys of homeowners have shown that comfort and aesthetic benefits far outweigh energy concerns. Very few homeowners assess the economic benefits of their investments by calculating payback times. Studies indicate top priorities in home upgrades are improving comfort and lowering operating costs.

FACTORS INFLUENCING SOLAR GAIN POTENTIAL

Heating degree days
A measure of the amount of time the temperature dips below 18°C
(68°F) on an annual basis

Length of heating season
Some regions have much shorter heating seasons than others, even at
the same latitudes

Number of sunlight hours in the coldest periods
The number of sunlight hours that reach your house during the coldest
times of the year determine the available solar gain

Angle of incidence to sun
The angle windows or collectors make with the sun determines
how much of the available solar gain you can access, and during
which seasons

Proximity to large body of water
Large bodies of water give coastal regions milder climates, but also can
lead to fog and longer periods of cloud

Latitude
Where you are on the globe determines the climate and the potential
solar gain

Altitude
Where you are in relation to sea level determines the severity of your
climate and the length of heating season

Climatic Zone
Cool climates range across ten global zones

Microclimate
The immediate surroundings determine the availability of the solar
gain potential, as do general temperature and weather patterns

The list below shows how some major cities rank in terms of
potential electricity generation from the sun.

City Yearly	PV potential (kWh/kW)
Cairo, Egypt	1635
Capetown, South Africa	1538
New Delhi, India	1523
Los Angeles, U.S.A	1485
Mexico City, Mexico	1425
Regina (SK), Canada	1361
Sydney, Australia	1343
Rio de Janeiro, Brazil	1253
Ottawa, Canada	1198
Beijing, China	1148
Washington DC, US	1133
Paris, France	938
Tokyo, Japan	885
Berlin, Germany	848
Moscow, Russia	803
London, England	728

Data for the PV ranging chart from www.canmetenergy-canmetenergie.nrcan-rncan.gc.ca/eng/renewables/standalone_pv/publications/2006046.html

SITE DESIGNING

The first step in any design process is to figure out what you have to work with. Understanding your 'raw materials' allows you to utilize their best qualities and compensate for their weaker aspects. In terms of a house, your raw materials are your building site and its relationship to the sun, the wind and the rain.

People sometimes fall in love with a house design, and then go looking for a place to build. If the house is not designed with that particular site in mind, it can lead to disappointment and frustration. A better approach is to first fall in love with a site or building lot. This allows you to design your house around what already exists.

If you are renovating your home or building an addition, an understanding of your site is also essential to making design decisions that improve the interaction between your existing house and the surrounding microclimate. This understanding also allows you to make full use of a building's solar potential and to plan for future solar installations.

Once you determine how the elements affect your site (climate, solar access, wind and water cycle), you begin to understand its character. Along the way, think about your lifestyle. Design is often a compromise between the site itself and your goals and purposes. Every site and every home are different. The ultimate goal is to achieve harmony between the setting you have and the home you want.

You exercise your creativity through the design process, you gain a deeper understanding of the place where you will be living, and you can optimize the solar contribution to your space and water heating through proper solar orientation and site design.

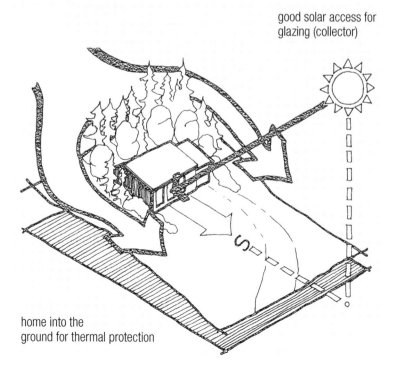

WINTER WINDS DEFLECTED

good solar access for glazing (collector)

S

home into the ground for thermal protection

site specifics

There are codes regulating everything from zoning to plumbing. Each will influence how you design your site and your house and you will have to abide by them throughout the building process.

The first restrictions you need to know about pertain to your site. A surveyor's map indicates your legal property lines. If your survey is extremely old, it may not be accurate. Consider having another one done. You may not be able to develop or build on a portion of your property because of easements or covenants put in place by prior owners. An easement gives someone else certain rights to a portion of your site. Easements can only be used for specific and limited purposes, and are often informal agreements between neighbours. To avoid any problems, ask the seller of the property and the neighbours about easements on your site. Common examples are utility poles which run across property lines, road access to another property, and access to views or sunlight from a neighbouring property. Covenants are private legal restrictions on the use of land, which are usually formally recorded on your property deed.

The municipal or regional planning office is the next stop for information that defines where and how you can build on

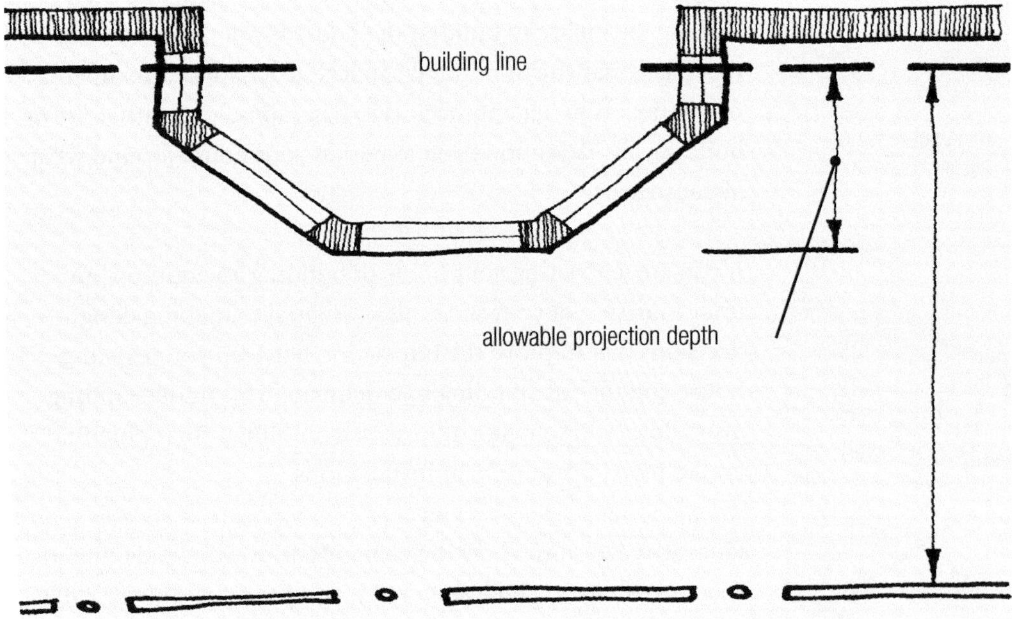

building line

allowable projection depth

In most areas, local building regulations are in place. It is a good idea to find out how they will affect your site and the surrounding area in the future.

HOW PROJECTIONS OR ENCROACHMENTS ARE MEASURED
The allowable projection depth is measured in metres, or as a percentage of the setback (or yard) depth.
Projections also have a maximum allowable width.
See your local building regulations.

your property. The planning office will inform you about zoning bylaws which affect your site, including the following common restrictions: setbacks (or yards), maximum building heights, floor area ratios and window exposures. There may be other restrictions that are applicable under the local zoning by-laws.

Setbacks are the required distance between your house and the boundaries of your site. They are required for many reasons: to provide emergency access; space for landscaping, parking, storage and recreation; visibility for traffic safety; noise reduction; views and light access. Setbacks apply to your house, but accessory buildings such as sheds and freestanding garages can be placed close to property lines in most jurisdictions. Some architectural features, such as bay windows, vestibules, eaves and decks are permitted to project into setbacks. How much these features are allowed to extend is shown as either a linear measurement or a percentage of the setback.

Maximum building height is usually measured from the highest point of the building to the ground, excluding chimneys, TV antennas, satellite dishes and other non-permanent projections. In most single-family residential zones, the maximum building height is somewhere around 10.7m (35ft), about the height of a two-storey house with a peaked roof. In

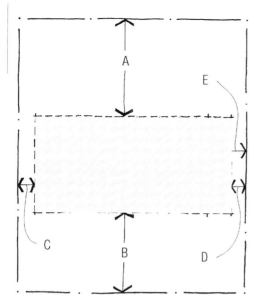

Front, rear and side yards or setbacks in plan.

TYPICAL MINIMUMS FOR SINGLE FAMILY RESIDENTIAL ZONES IN NORTH AMERICA

2–10 acres

60–120m (200–400 ft) frontage

6m (20ft) front yard (A)

7.6m (25ft) rear yard (B)

Urban lots: 370–465m² (4,000-5,000ft²)

12–15m (40–50ft) frontage

4.6–6m (15–20ft) front yard (A)

7.6–10.6m (25–35ft) rear yard (B)

Side yard minimums for any residential zone (C,D,E):

(C,D) 1.2m (4ft) plus 0.6m (2ft) for each storey over 1, both sides if you are planning to have an attached garage.

(E) If you are planning to have a freestanding carport or garage, then one side yard must be 3–3.7m (10–12ft) to allow for driveway width.

some cases, special rules cover structures built on steep slopes or close to open water, or both. Your local bylaws may also indicate a maximum allowable height for fencing in your area.

The floor area ratio (FAR) limits the size of your house. It is the ratio of floor area permitted on the site to the size of the site. The gross floor area is the sum of all the floors of a structure measured from the outside surfaces of exterior walls. In most regions the following features or areas are excluded: attached garages, porches, verandahs, sunrooms, balconies, unfinished attics, basements and cellars. Make sure you know what is pertinent to your zoning or planning district.

If, after outlining your setbacks and other restrictions, you find that you are unable to build on your site, you will have to go through a formal appeal process to get a variance to the bylaw affecting your site. Staff at the planning office can tell you how to go about doing this. A common example of a variance is to reduce setbacks because the lot shape or topography makes it impossible to build within those required.

If you live in an area not serviced by municipal water mains or sewers, a local health inspector must assess and design your disposal system early in the site design process. The disposal field must be placed where it will function properly, and where it won't contaminate your well. So, to some extent, it will dictate where you can place your house on your site.

SITE RESTRICTIONS

Urban and controlled subdivision sites will be subject to more stringent zoning bylaws and regulation than rural sites because of the smaller lot sizes and higher density levels. With this in mind, here is a 'typical' small urban lot and the most common zoning bylaws that will affect it. Corner lots may be subject to a 'corner vision triangle', to allow for clear and safe visibility at intersections. This means you cannot build within a triangle of specified dimension at an intersection. Also, fences, signs, hedges and shrubs must be kept below a certain height within that triangle. Any restrictions will be outlined in your local bylaws.

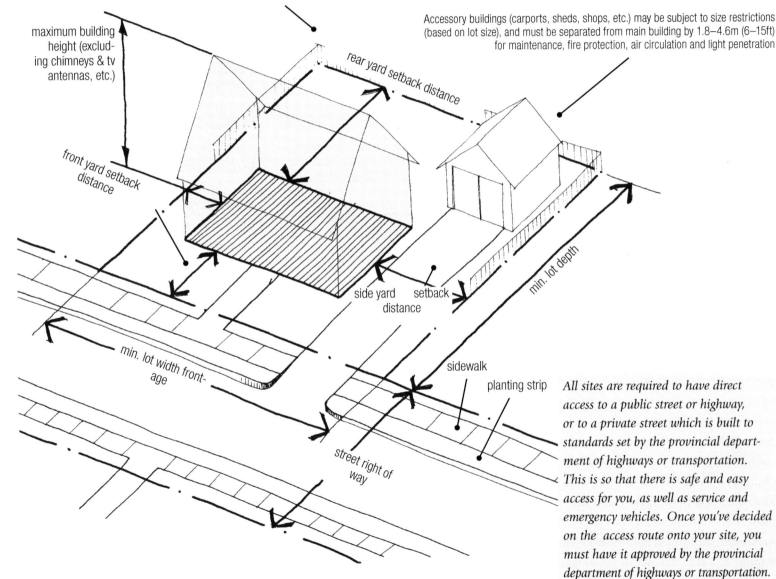

fencing restrictions may apply

Accessory buildings (carports, sheds, shops, etc.) may be subject to size restrictions (based on lot size), and must be separated from main building by 1.8–4.6m (6–15ft) for maintenance, fire protection, air circulation and light penetration

maximum building height (excluding chimneys & tv antennas, etc.)

rear yard setback distance

front yard setback distance

side yard distance

setback

min. lot depth

min. lot width frontage

sidewalk

planting strip

street right of way

There are also regulations pertaining to open watercourses such as rivers and streams, building on oceanfront and other shorelines. Usually, the regulations state that you cannot build within a certain distance of the high water mark of any watercourse or ocean. In some cases, you can infill or alter shorelines, but you must have your plans approved by one or more federal or provincial agencies (such as departments of environment, natural resources and fisheries and oceans).

All sites are required to have direct access to a public street or highway, or to a private street which is built to standards set by the provincial department of highways or transportation. This is so that there is safe and easy access for you, as well as service and emergency vehicles. Once you've decided on the access route onto your site, you must have it approved by the provincial department of highways or transportation.

natural cycles

As a solar homeowner, you must be aware of the annual solar cycle, and how it will affect your site and your home. The sun rises in the east, reaches its highest point at noon, and sets in the west. When and where the sun rises, how high it climbs in the sky and when and where it sets all depend on the time of year.

When designing your solar house, the position of the sun at the winter and summer solstices determines certain design parameters. In cool climates, the prime concern is to keep warm in the winter. Summer cooling is important in

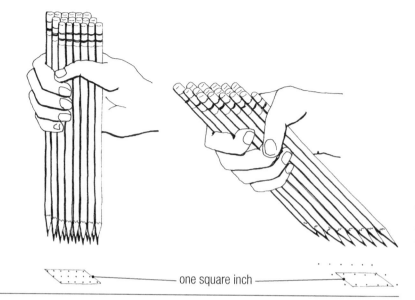

— one square inch —

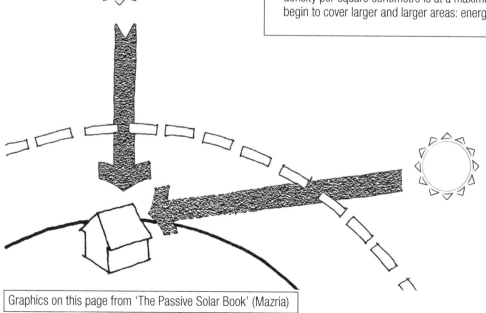

ANGLE OF INCIDENCE

Since the radiation coming to Earth is in essentially parallel rays, surfaces perpendicular to those rays intercept the greatest amount of energy. As the sun's rays move away from being perpendicular, the energy intercepted by a surface decreases.

One way to imagine this is to think of the parallel rays of the sun as a handful of pencils held with their points touching a horizontal surface. The dots made by the points represent units of energy. When the pencils are held perpendicular to the surface, the dots are as compactly arranged as possible: energy density per square centimetre is at a maximum. As the pencils are inclined towards the parallel, the dots begin to cover larger and larger areas: energy density per square centimetre is decreasing.

ATMOSPHERIC DISSIPATION

The length of atmosphere the sun's rays must pass through affects the intensity of the radiation received by the Earth. On a daily basis, when the sun is at its apex (noon), the sun's rays travel through the least amount of atmosphere. As the sun moves closer to the horizon (sunset), the path of the sun's rays through the atmosphere lengthens. The more atmosphere the radiation must pass through, the less its energy content due to increased absorption and scattering in the atmosphere.

Because of the Earth's tilt and rotation, the length of atmosphere radiation passes through varies not only with the time of day, but with the month of the year as well.

The Earth's elliptical path and the tilt of the Earth's axis combine to create the changing angles at which the sun's rays hit the surface.

Here is how to calculate the approximate sun angles for your latitude:

EQUINOX (spring or fall): The sun is over the equator (because declination due to the tilt of the Earth is zero).

$90° -$ your latitude = sun's height at noon

SUMMER SOLSTICE: The sun is over the Tropic of Cancer (23.5°N). Highest noon height above the horizon.

$90° + 23.5° -$ your latitude = sun's height at noon

WINTER SOLSTICE: The sun is over the Tropic of Capricorn (23.5°S). Lowest noon height above the horizon.

$90° - 23.5° -$ your latitude = sun's height at noon

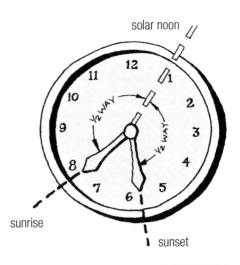

solar noon

sunrise

sunset

some areas, but in others the summer is not too hot and can be very brief. A well-designed solar home takes care of cooling passively in most regions, so your site and home design should be geared to winter conditions.

The winter solstice (around December 21) is the shortest day of the year, when the sun stays closest to the horizon and rises and sets south of the equator. The winter solstice determines the minimum hours of solar gain. At this time the intensity of the sun's energy reaching the Earth's surface is at its lowest. Your home needs to be designed to collect and store as much of this winter solar gain as possible.

The summer solstice (around June 21) is the longest day of the year, when the sun reaches its highest point in the sky, and rises and sets north of the equator. The summer solstice determines the maximum hours of solar gain, as well as any shading requirements and the limits of possible glazing angles needed to avoid overheating.

The angle at which the sun's rays strike the Earth varies on both a daily and seasonal basis. These variations are a result of two factors, and are very important in determining how much of the sun's energy is available to us, and at what times.

FINDING TRUE SOUTH
The needle of your compass points to magnetic north. However, there is a difference between magnetic north and true north. This is caused by variations in the magnetic field of the Earth. The variation differs with location and is called magnetic declination, see Appendix B. Your compass can be thrown off by the presence of metal, including your belt buckle. To avoid any misreadings, and to simplify the process, use the time method, outlined at right.

HOW TO FIND SOUTH

If you check the newspaper, TV weather channel or call your local environment agency, you can find the times of sunrise and sunset for a given day. From this, you can determine how many hours of daylight there will be. The sun will be at its highest, or true south, at the half-way point of its journey across the sky. By adding half of the daylight hours to the time of sunrise, you will know the exact time of day the sun is highest. If you are facing the sun at this time, you will be facing true south. This method corrects for Daylight Saving Time. To keep a 'permanent' record of true south on your site, do the following: put a stake into the ground (it must be absolutely perpendicular to the ground). At solar noon, place another stake several feet along the shadow cast by the first stake. East and west are laid out at right angles to the north-south line of your stakes.

EG: 7:45am to 5:45pm = 10hr/2 = 5hr
Solar Noon is: 7:45am + 5hr = 12:45pm

If you imagine a clock face on the ground, with 12 o'clock as south, you can figure out where your optimal solar orientation is, which direction the ground is sloping and if trees or other buildings will obstruct your solar gain. The numbers on the 'sundial' do not correspond to the actual time; they indicate direction in the same way that the military uses the clockface (i.e., incoming plane at ten o'clock).

The sun rises and sets north of east and west in the summer, and south of east and west in the winter. Note also that the sun is at the same place at the same hour all year round, but its height changes. This is important to know, because it tells you if the area you are in has good solar exposure through the winter, when the sun is lower in the sky.

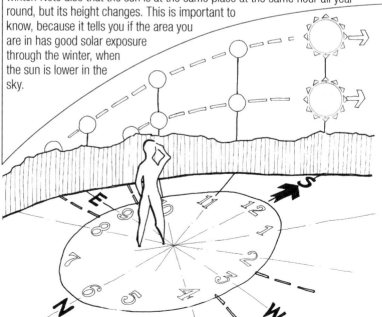

First, the angle the sun's rays make with a surface determine how much energy that surface receives. This is known as the 'angle of incidence'. In the summer, the sun's rays strike the Earth at an angle close to perpendicular. This means more energy reaches the Earth in the summer than in the winter, when the sun travels low across the horizon, and the sun's rays and the Earth's surface are closer to parallel. (See illustration on page 11 for detailed explanation.)

Second, the amount of atmosphere (air mass) the sun's rays have to pass through before reaching the Earth determines the intensity of the direct sunlight. The more atmosphere the sun's rays pass through, the less intense the radiation that reaches the Earth. This is because clouds, dust and other particles in the upper atmosphere absorb and scatter the sun's energy.

In orienting your house to the sun, you want the bulk of your windows, or other solar collectors, facing south. Facing true south is ideal but facing up to 30 degrees either west or east will have a negligible effect on your solar gain. If you are more than 30° east or west of south, you will begin to lose more heat than you gain in winter through double-glazed windows. The more severe your winter climate, the more important it is to face due south, because you will want to get the maximum amount of solar gain possible.

The Earth turns one revolution, 360°, in 24 hours. This means it spins at a rate of 15° longitude every hour (360 ÷ 24). Each of the numbers on a 12 hour clock are set at 30° intervals around the circle. Each of these intervals represents two hours of the sun's path across the sky.

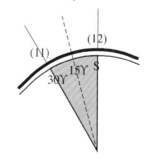

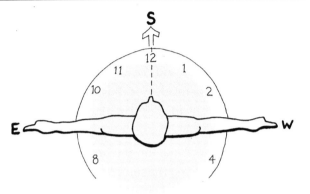

Energy from the sun reaches the Earth's surface in three ways. Direct radiation is the most intense. When direct radiation is scattered by dust, air molecules and clouds, it becomes diffuse radiation. Reflected radiation is direct or diffuse radiation bounced off surfaces such as water or snow.

direct radiation

diffuse radiation

reflected radiation

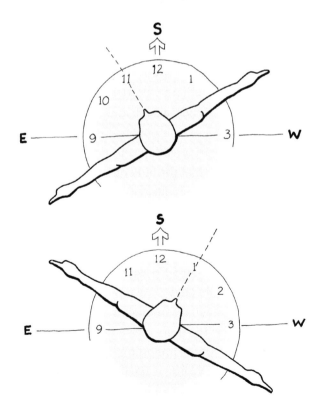

Orienting your house off south will affect the time of day when you will get your best solar gain. Facing south, your best gain is at noon. For every 15° off south, you change the time by one hour. So, if you are facing 15° east, you will get your highest amount of solar gain at about 11:00 in the morning. If you are 30° west, you will get your maximum solar gain at about 2:00 in the afternoon. In the summer, this means that you will be getting your heat gain during the hottest part of the day, with a good possibility of overheating.

We tend to spend more time outdoors during the fine weather of summer. When designing outdoor spaces, think about the times when you will want to use them. For instance, if you work all day and are home in the evening, it might be best to plan your outdoor space on the west side of your house where the setting sun will continue to provide heat. If you spend your days at home and want to be outdoors midday, consider a northern or eastern location so that direct southern exposure does not overheat your area and make it uncomfortable.

You can create outdoor 'suntraps' by carefully housing the space from wind. If this is done without limiting the solar gain to the housed area, then you can raise the average temperature slightly, so that the area is comfortable earlier in spring and later into the fall.

weather and climatic regions

Like the seasons, changes in the weather come in predictable cycles, but these changes lag behind the solstices and equinoxes by a month or two. This lag is accounted for by the fact that it takes time for the air to warm up in spring or cool down in the fall.

The 'climate' of a region is determined by long-term weather data,

which shows the characteristic condition of the atmosphere near the Earth's surface. While the seasonal weather changes are dictated by global trends, there is an enormous amount of variability across any region.

There are three climate groups, and each of these groups is broken into subgroups. Group I are the low-latitude warm climates, controlled by equatorial and tropical air masses.

CLIMATE REGIONS

GROUP II MID-LATITUDE CLIMATES

steppe

Characterized by grasslands, this is a semiarid climate.
Moist ocean air masses are blocked by mountain ranges to the west and south.
These mountain ranges also trap polar air in winter, making winters very cold.
Summers are warm to hot.
Temperature Range: 24°C (43°F).
Annual Precipitation: less than 10cm (4in) in the driest regions to 50cm (20in) in the moister steppes.
Latitude Range: 35°–55° N.
Extent: Western North America (Great Basin, Columbia Plateau, Great Plains); Eurasian interior, from steppes of eastern Europe to the Gobi Desert and North China.

grasslands

Similar to the steppe, but with more precipitation. In the summer, a local continental air mass is dominant.
A small amount of rain falls during this season.
Annual temperatures range widely.
Temperature Range: 31°C (56°F).
Annual Precipitation: 81cm. (32in.).
Latitude Range: 30°–55° N and S
Extent: North America (Great Basin, Columbia Plateau, Great Plains); Eurasian interior, deciduous forest

The 'polar front' zone. Seasonal changes are very large. Daily temperatures change frequently. High levels of precipitation year-round, increased amounts in the summer due to tropical air masses. South-moving polar and arctic masses create cold winters.
Temperature Range: 31°C (56°F)
Average Annual Precipitation: 81 cm (32 in).
Latitude Range: 30°–55° N and S (Europe: 45°–60° N).
Extent: eastern parts of the United States and southern Canada; northern China; Korea; Japan; central and eastern Europe.

GROUP III HIGH-LATITUDE CLIMATES

taiga

Also known as the boreal forest, with long, very cold winters, and short, cool summers. This continental climate in the polar air mass region has a wider temperature than any other climate.
Precipitation increases during summer months, although annual precipitation is still small.
Temperature Range: 41°C (74°F), lows; -25°C (-14°F), highs; 16°C (60°F).
Average Annual Precipitation: 31cm (12in).
Latitude Range: 50°–70° N and S.
Extent: central and western Alaska; Canada, from the Yukon Territory to Labrador; Eurasia, from northern Europe across all of Siberia to the Pacific Ocean.

tundra

Polar and arctic air masses dominate the tundra climate, along arctic coastal areas. The winter season is long and severe.
A short, mild season exists, but not a true summer season.
Moderating ocean winds keep the temperatures from being as severe as interior regions.
Temperature Range: -22°C to 6°C (-10°F to 41°F).
Average Annual Precipitation: 20cm (8in).
Latitude Range: 60°–75° N.
Global Position: arctic zone of North America; Hudson Bay region; Greenland coast; northern Siberia bordering the Arctic Ocean.

alpine

Alpine, or highland, climates are cool to cold, found in mountains and high plateaus. Climates change rapidly on mountains, becoming colder the higher the altitude gets. Apline areas have the same seasons and wet and dry periods as the broader climatic region they are in, although the climate becomes colder with increased altitude. Alpine climates are very important as seasonal water storage areas.
Temperature Range: -18°C to 10°C (-2°F to 50°F)
Average Annual Precipitation: 23cm (9in.)
Latitude Range: found all over the world
Extent: Rocky Mountain Range in North America, the Andean mountain range in South America, the Alps in Europe, Mt. Kilimanjaro in Africa, the Himalayas in Tibet, Mt. Fuji in Japan.

For detailed information about your area, use keywords: climatic zones, plant hardiness maps.

Group II, the mid-latitude climates, are affected by both tropical and polar air masses that are in nearly constant conflict. Three of the four subgroups are classed as cool climates (the fourth – mediterranean – can still benefit from cool climate solar strategies). Group III, the high latitude climates, are dominated by polar and arctic air masses. All three subgroups are classed as cool climates.

In terms of designing a solar home, latitude plays a big part in what you can expect to gain from the sun. Planning a passive solar house at latitude 40°N is very different than planning one at latitude 70°N. Sun hours and sun angles influence everything from sizing windows, solar collectors and overhangs for shading. Decisions also need to be made about what kind of technologies will give the best performance in different seasons.

sun cycle/temperature cycle

The shortest day of the year is around December 21st, but the coldest time of the year is typically mid-February. On the other side of the year, the longest day is around June 21st, but the hottest time of the year is typically mid-August. Thus, the temperature cycle follows the sun cycle by about eight weeks, creating a 'thermal flywheel' effect.

Passive solar designs maximize the winter heat gain (easy to do) and minimize summer heat gain (not so easy to do). The lag between the sun cycle and the temperature cycle is beneficial to maximizing heat gain in the winter, but causes significant challenges to minimizing summer heat gain. By mid-August, the sun cycle is much lower in the sky, increasing the solar gain during the hottest time of the year – those long, hot, lazy summer afternoons that we all dream of in February. This is the time of year when passive solar homes overheat because the lower sun increases the gain on south-facing vertical glass. Orienting

off south increases the effect, especially if you orient the glass to the west, maximizing the heat gain at the hottest time of the day and the year.

For active solar, the sun/temperature cycle flywheel means that come mid-February, the stronger sun and the longer day increase the performance of these systems.

THE FLYWHEEL EFFECT
The Earth's seasonal lag can be used to moderate the temperature extremes experienced by your house, as shown by the chart below.

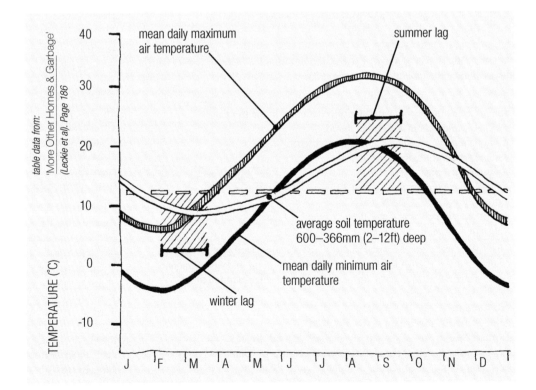

winds

Since strong winds increase the infiltration of air into your house and accelerate the conduction of heat through glazing and walls, the potential for heat loss on a cold windy day is much greater than on a calm day. You can defend against cold winds by retaining lots of tree cover on the north side of the house. Trees will break the force of the wind or deflect it over the house, reducing heat loss. If you have no tree cover, you can plant a windbreak that will need to be established for a few years before becoming truly effective.

Similarly, if you find that summer breezes tend to come from one direction, you can try to use them to your advantage. Channeling summer breezes through trees or around berms can keep your house cool and mosquitoes at bay. You can take advantage of natural ventilation and cooling in the summer by placing windows or vents in strategic areas. This will be covered in the house design and climate control sections.

Remember too, that cold air falls and warm air rises. This is why low-lying areas tend to get early frosts. In siting your house, try to avoid placing it in a depression or in an area that does not get sun until late in the day. If you are planning to build on a slope, the best placement is

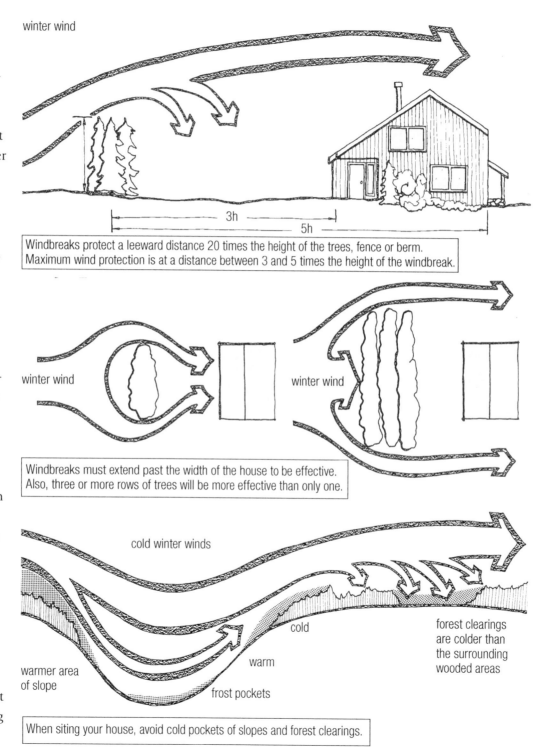

winter wind

3h

5h

Windbreaks protect a leeward distance 20 times the height of the trees, fence or berm. Maximum wind protection is at a distance between 3 and 5 times the height of the windbreak.

winter wind

winter wind

Windbreaks must extend past the width of the house to be effective. Also, three or more rows of trees will be more effective than only one.

cold winter winds

cold

warm

warmer area of slope

frost pockets

forest clearings are colder than the surrounding wooded areas

When siting your house, avoid cold pockets of slopes and forest clearings.

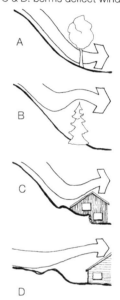

A: wind travels through deciduous trees
B: wind is deflected over low-branched conifers
C & D: berms deflect wind

A

B

C

D

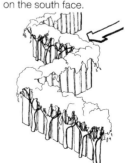

'FEDGE'
(FENCE HEDGE)
A zig-zag pattern is more stable than a straight fence. Plant evergreen vines like climbing hydrangea or euyonomous (winter creeper) on the south face.

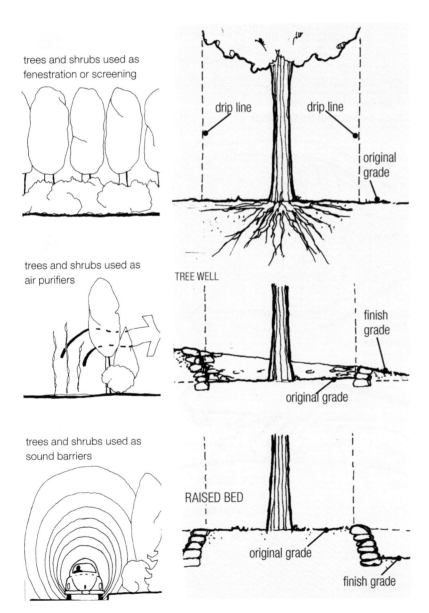

trees and shrubs used as fenestration or screening

trees and shrubs used as air purifiers

trees and shrubs used as sound barriers

TREE WELL

drip line   drip line

original grade

finish grade

original grade

RAISED BED

original grade

finish grade

PRESERVING TREES ON YOUR SITE

Trees, like humans, must breathe. The main way they do this is through their leaves. Another way is through the bark. If you must change the grade of your site to accommodate your house, but want to save a single tree or a cluster of trees, you must allow them to continue to breathe through their bark. To protect the tree, the soil needs to stay at the original level. You can build either a tree well (left centre) or a raised bed (left bottom).

These drawings show the general configuration of tree wells and raised beds. They should extend to the 'drip line' of the tree, as the circumference of the branches and leaves roughly reflects the circumference of the root area. Different species require different treatment. Talk to your local nursery or an arborist for specifics.

somewhere down the middle of the hill. This allows you to use the top part of the slope as a buffer from the prevailing wind, without getting into the cold pocket at the base of the slope (see illustration – page 17).

With higher interest in wind energy, wind patterns are being mapped for more areas. These maps show a wind 'rose', which indicates the prevailing winter and summer winds in an area, as well as average and peak seasonal wind speeds. These can be helpful in determining how to minimize the impact of cold winter winds and how to take advantage of cooling summer breezes.

trees

Trees are dynamic. They grow and they die. Consider this when you plant or design around small trees near your house. They will grow into large trees and may over-shade your roof and south-facing walls, and by then, you may have developed a fondness for them. You may be loathe to cut them down to improve your solar gain. Better to avoid trees on the south near to the house. Instead, plant native or hardy shrubs that will grow only a few feet high.

Also, many people believe that it is beneficial to have hardwood trees on the south side of the house, assuming that the trees will shade the house in summer and allow the sun in during winter, when the leaves have fallen. The fact is that the branches of bare trees can block up to 60 per cent of the sunlight falling on your house, depending on the species of tree.

An important consideration on some sites is how much of the tree cover to re-move. Most coniferous trees are shallow-rooted. If you weaken the stand's wind resistance by removing some trees, you may experience windfalls in the future. .If it can't be avoided, clear the trees far enough away from the house so that if they fall they won't damage the house.

SUCCESSION:
In the constant renewal of the landscape, pioneer species show up after events such as fires or logging. These quick-growing species stabilize drainage patterns, reduce erosion, and improve soil quality for later comers: slow-growing species, or species which require shaded areas to grow. A climax forest is one in which a large number of trees of both fast and slow growing species have reached maturity. A climax forest can remain stable for many years. A successional forest is one in which many of the the species have become overmature and begin to die out. At this stage, it may be wise to cut down the most decayed trees to let the young plants that succeed them have access to sunlight.

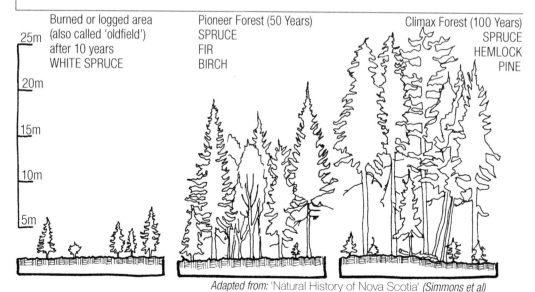

Burned or logged area (also called 'oldfield') after 10 years
WHITE SPRUCE

Pioneer Forest (50 Years)
SPRUCE
FIR
BIRCH

Climax Forest (100 Years)
SPRUCE
HEMLOCK
PINE

25m
20m
15m
10m
5m

Adapted from: 'Natural History of Nova Scotia' (Simmons et al)

As a rule of thumb, any object (tree, building, etc.) south of your solar house should be no higher than half its distance away or it will shade the glazing in midwinter. Remember, though, that trees grow taller!

MEASURING THE HEIGHT OF TREES AND OTHER TALL THINGS

Have your partner stand at the base of the tree with a measuring tape. You stand far enough away from the tree so that you can close one eye and see the tree from top to bottom along the length of a pencil. The top of the tree should be at the end of the pencil, and the tip of your thumb should be at the bottom (A). Turn the pencil on its side, keeping your thumb at the centre line of the tree's trunk and the pencil at eye level. Your partner then takes the measuring tape, with one end secured at the centre line of the trunk, and walks away from the tree (at a right angle to you) until they reach the end of your pencil. The tree's height is the length measured off by your partner's measuring tape (B).

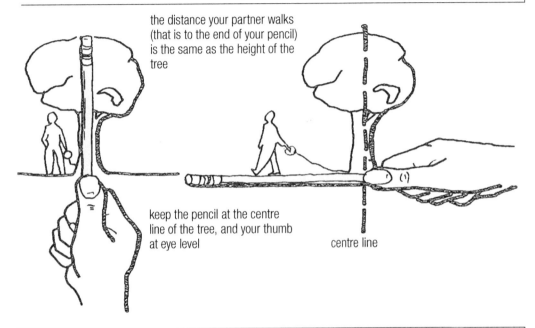

the distance your partner walks (that is to the end of your pencil) is the same as the height of the tree

keep the pencil at the centre line of the tree, and your thumb at eye level

centre line

The further north you go from 45°N, the less accurate the half-height rule becomes. At latitude 60°N, trees or buildings within 3x their height would be considered obstructions for spring and fall. Mid-winter sun is so low in the sky at northern latitudes that anything closer than 4x the height could cast a shadow on the building at noon.

Full-grown trees or other tall things to the south of your home should be no higher than half their distance from the major collector surface of your house.
The half-height rule doesn't account for slope on your site, but will give you a rough estimate on all but very steep sites.

HALF HEIGHT RULE

Now that you know the height of trees, existing buildings or other tall things on or near your site, and you know where true south is, you can determine which objects will be obstructions; that is, which ones will shade your house during the critical hours of midday, or 30° either side of true south. Anything that doesn't comply with the 'half-height' rule, illustrated below, will be an obstruction.

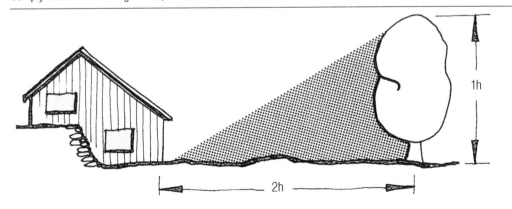

1h

2h

microclimates

Microclimates are areas that are constantly a few degrees higher or lower than surrounding areas. They are caused by combinations of wind protection, topography and exposure to solar gain. Microclimates can be detected by observing any differences in the times that similar plants bud out in the spring or die back in the fall. The type of plant and tree cover also gives you visual indications of the local soil and weather conditions.

Careful analysis of microclimates on your site can be done whether you are in an inner city neighbourhood or a big rural acreage. Do neighbouring high-rise buildings create a wind tunnel or do they block the prevailing winter winds? Will that south-facing ridge be well-protected from the winter winds? You want to avoid placing your house in areas which are cooler, and take advantage of the 'heat traps' that exist on your site.

measuring slopes

Most sites have a certain amount of slope, or grade. Slopes can be measured in proportions (as a rise to run ratio), in degrees from the horizon, or as percentages (especially in engineering). The easiest expression to use on site is the rise to run ratio, as you can easily convert it in your head or with a simple hand calculator.

You need to know the range of the slopes on your site for several reasons: placement of the disposal field and well or connection points for municipal sewer and water mains, placement of your house and access routes. Slope orientations also determine the amount of possible solar gain, where you place your windows or other solar collectors, and the degree of protection from winter winds.

As explained earlier in this section, slopes also influence drainage patterns, which are influenced by soil conditions. These three elements – slopes, drainage and soil conditions – are interrelated, and should never be looked at in isolation from each other in the overall site design process.

The optimum slope in most cold climates has an easy grade and a south or southeast orientation. A southeast slope has the advantage of exposure to

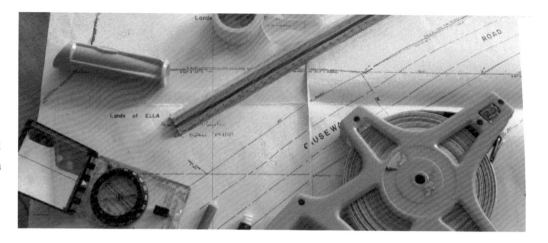

MEASURING SLOPES
You need two measuring tapes, stakes, string and a bubble level that can be hooked onto the string. First, measure off some distances parallel to the orientation of the slope. If you can see where a slope changes (where it becomes steeper for example), then break that measurement into two lengths, as you need to know the pitch of each slope. Stake or flag each length that you measure. Once you have some measured lengths, tie the string to the base of the uphill stake and run it to the downhill stake, but don't tie it off. Place the bubble level on the string (downhill end) and move it up or down until the level shows that the string is horizontal. The 'rise' of the slope is measured by holding a tape measure vertically and recording the height of the string from ground level. If you are in thick underbrush, tie the string higher on the uphill stake. Don't forget to subtract the string's height above the ground (on the uphill stake) from your downhill vertical measurement.

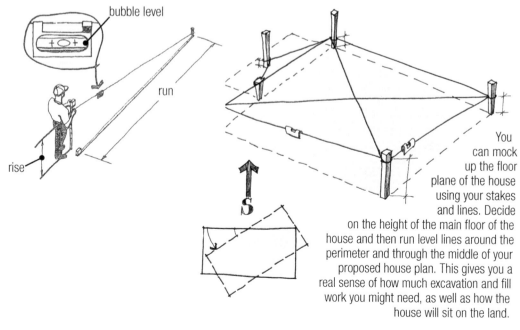

bubble level

run

rise

You can mock up the floor plane of the house using your stakes and lines. Decide on the height of the main floor of the house and then run level lines around the perimeter and through the middle of your proposed house plan. This gives you a real sense of how much excavation and fill work you might need, as well as how the house will sit on the land.

USING A HAND LEVEL
Record the height at which you hold the level (your eye level). Your partner stands at a measured distance with a vertical measuring tape. The measurement is indicated by crosshairs in the sight tube. The rise of the slope is the difference between your eye level and the crosshair height.

Some common slopes	
slope %	ratio
0.5	1:200
1	1:100
2	1:50
2.5	1:40
3	1:33.33
4	1:25
5	1:20
6	1:16.67
7	1:14.28
8	1:12.50
8.33	1:12
9	1:11.11
10	1:10
20	1:5
33.33	1:3
50	1:2

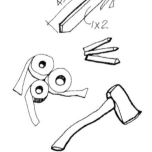

Have a good supply of flagging tape or brightly coloured rags, lumber, crayons or magic markers, and 1 x 2 stakes.
A hatchet will knock stakes into the ground, mark trees to be cut and clear underbrush.

HOW TO DETERMINE THE SLOPE PROPORTION

The height difference between two points is the rise of the slope, while the horizontal distance between those two points is the run.

SLOPE = RISE ÷ RUN   RISE = SLOPE × RUN   RUN = RISE ÷ SLOPE

Here's a metric example: The vertical change is 20cm. The distance between the two measurements is 6m. This is 20cm of rise to (6 x 100cm) of run, or a ratio of 20:600. Dividing this figure by 20 (20:600/20) gives a slope of 1:30.

Here's an imperial example: The vertical change is 6 inches. The distance between the two measurements is 20ft. This is 6 inches of rise to 20×12 inches of run, or a ratio of 6:240. Dividing this figure by 6 (6:240/6) gives a slope of 1:40.

Slope is proportional, and so has no unit of measurement. In the examples, the slope 1:40 indicates that for every 1 unit of rise, you have 40 units of run. A percentage expresses the number of units of rise in 100 units of run. For example: 3 per cent indicates 3ft of rise in 100ft of run, or 1:33.

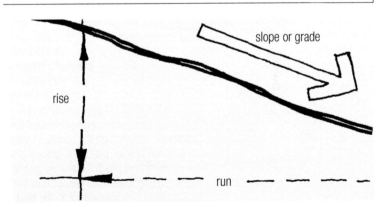

summer breezes, but summer cooling through orientation is not as important in some regions as solar access and wind protection in the winter. We will deal with orientation and solar gain in more depth in the house design section. Once you have determined the orientations of the slopes on your site, you must determine how steep they are.

When measuring slope, you determine the difference in the height (rise) of the land over a specified distance (run). This can be done using such simple tools as string or a bubble level and a measuring tape. These tools are reasonably accurate: your slope calculations will be close enough to lay out your house, access roads, and any earthworks which don't require engineering expertise. To be able to accurately determine the slope, you need to create a flat plane. First, pound in a set of posts marking out the rough estimation of the main corners of your house. Then, using your bubble level and lengths of string tied to each post, create the flat plane. You might find your site slopes in more than one direction, or that it is much steeper than you think!

So now that you know how to measure the slopes on your site, what can you do with the information you have gathered? Most aspects of site and house design are limited by, or dependent on, slopes. There are maximum slopes for stable building foundations, maximum slopes that cars and excavation equipment can handle, and maximum slopes that people can walk up or down comfortably. Every soil type has an 'angle of repose', that is, the slope at which it is no longer stable. This is important when building a house into a slope, when constructing earthworks,

and when building roads. Otherwise, you will have erosion and potentially dangerous areas on your site.
See Appendix for more on slopes and earthworks.

- Flat areas with a slope under 1:100 do not drain well, and should be avoided for large disposal fields and building sites. They may, however, work well as access roads or driveways, provided they are carefully finished and have adequate drainage installed.
- Slopes under 1:25 can be used for disposal fields, building sites and roads without too much difficulty on most soils if the drainage is adequate.
- Slopes over 1:10 can be more expensive and complicated to build on, but there are advantages to them: building your house into a steep slope can protect it from cold winter winds, and you can enjoy areas such as roof terraces.

Soil conditions and drainage patterns will dictate how you can use any of these areas. When working with steep slopes, the bearing capacity of a soil becomes a limiting factor. Steep slopes may also require specially designed disposal fields or sewage and water main connections. Even if the slopes on your site are not steep, placing the long axis of your house parallel to the contours of the slope can reduce the expense of construction, grading and filling. When you are on site and measuring slopes, you can stake out the approximate area of your building, with the long axis across the contour to get an idea of how it could sit on the site.

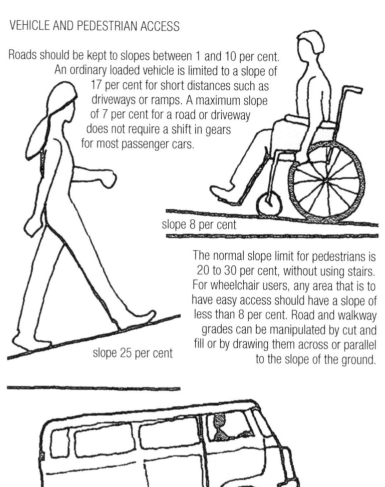

VEHICLE AND PEDESTRIAN ACCESS

Roads should be kept to slopes between 1 and 10 per cent. An ordinary loaded vehicle is limited to a slope of 17 per cent for short distances such as driveways or ramps. A maximum slope of 7 per cent for a road or driveway does not require a shift in gears for most passenger cars.

slope 8 per cent

The normal slope limit for pedestrians is 20 to 30 per cent, without using stairs. For wheelchair users, any area that is to have easy access should have a slope of less than 8 per cent. Road and walkway grades can be manipulated by cut and fill or by drawing them across or parallel to the slope of the ground.

slope 25 per cent

slope 10 per cent

slope 17 per cent

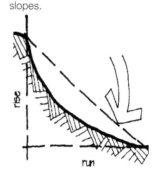

SLUMPING SLOPES
Concave slopes duplicate the natural 'slump' of the soil and are more stable than straight slopes.

rise

run

ANGLES OF REPOSE
Recommended proportions of slopes for stable angles of repose over a range of soil materials.

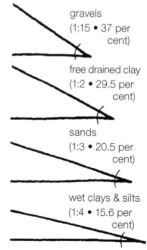

gravels
(1:15 • 37 per cent)

free drained clay
(1:2 • 29.5 per cent)

sands
(1:3 • 20.5 per cent)

wet clays & silts
(1:4 • 15.6 per cent)

from 'Permaculture', A Practical Guide for a Sustainable Future *page 230* (Bill Mollison)

There are several ways to surface your access routes besides the typical concrete or asphalt. Paving blocks of all types and sizes are available. Gravel and shale are available in most regions. Avoid asphalt and asphalt derivative surfaces, as they tend to leach contaminants into groundwater sources.

Whatever surface you choose, first make sure that the roadbed is well-graded and will drain properly. Be aware of any trees or steep banks which shade the roadbed in winter. They will increase the amount of ice (and longevity of ice) that could form on the road.

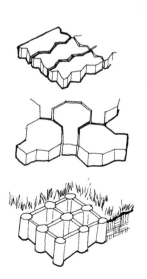

DRIVEWAYS & PARKING

- increase driveway widths 0.3m (1ft) at curves, add 0.3m (1ft) for walkway
- length of parking space can be reduced to 4.9m (16ft) for a small car
- large trucks may require a larger parking and turnaround space
- add 1m (3.3ft) to width of carport enclosing an entrance to the house
- parking for disabled persons must be at least 5m (16ft) wide, for easy maneuvering of chairs and other mobility aids

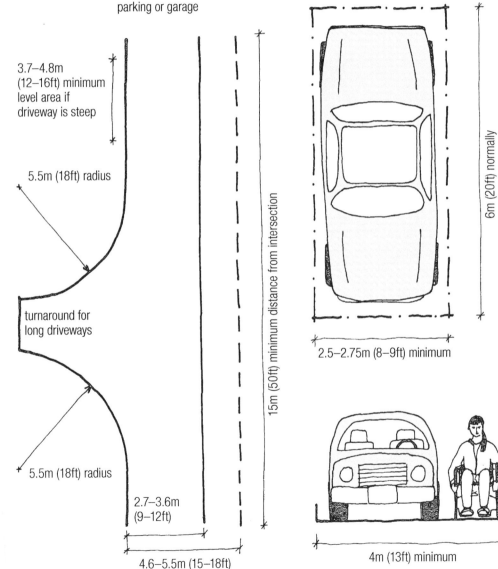

parking or garage

3.7–4.8m (12–16ft) minimum level area if driveway is steep

5.5m (18ft) radius

turnaround for long driveways

5.5m (18ft) radius

2.7–3.6m (9–12ft)

4.6–5.5m (15–18ft)

15m (50ft) minimum distance from intersection

2.5–2.75m (8–9ft) minimum

6m (20ft) normally

4m (13ft) minimum

GUIDELINES FOR SOME IMPORTANT LENGTHS & SPACINGS
(These may vary in your region.)

- *Max. distance from supply or emergency vehicle to door: 75m (250ft)*
- *Min. separation between driveway entrance and intersection: 15m (50ft)*
- *Width of entrance walk: 0.8m (2.5ft)*
- *Min. width of driveway: 2.5m (8ft)*
- *Min. width of road clearance for dump trucks, etc.: 5.5m (18ft)*
- *Min. width of a right–of–way negotiable by a small vehicle: 3m (10ft)*
- *Parking length: 6m (20ft)*
- *Parking width: 2.5 to 2.75m (8–9ft)*
- *Disabled parking width: 2.5 × 5m (8 × 16ft)*

Other Access:

- *Max. length of run, low voltage line: 120m (400ft)*
- *Normal spacing, electric power poles: 40m (120ft)*
- *Max. reach, oil truck hose: 30 to 60m (100 to 200ft)*

Lighting:

- *Standard height street lights: 9m (30ft)*
- *Spacing between street lights: 45 to 60m (150 to 200ft)*
- *Mounting heights, walkway lights: 3.5m (12ft) or lower*

site access

There are several different ways in which your site needs to be accessed. The most obvious is an access road or a driveway for your vehicles. What about people on bikes, or pedestrians? You can make paths, boardwalks and bridges. Utilities – municipal water and sewer lines, electricity, telephone/cable TV – need to have their own access routes. Emergency vehicles need to be able to access your site easily.

Imagine approaching and arriving at your site. What do you see first? Do you see your house from the street or part way down the drive? Should your drive come off the street squarely or at an angle? What part of the house do you see first? Is the entry obvious? Are your vehicles visible or hidden? Do you have space to back and turn them? How do you get to the house? Is your entry housed? Do you have to go up stairs? How do you remove snow? What about access to other outdoor activities?

Try outlining your driveway and outdoor spaces with stakes and visible twine to get a feeling for their contours. Make the best use of what the site has to offer. Imagine landscaping, built elements and finishes. A paved driveway will create a different effect from a gravel driveway, for instance.

You should lay out your access roads in terms of the worst driving conditions you will experience: winter's snow and ice, the spring thaw, heavy rainstorms and the like. Think about how your vehicle performs in bad winter conditions and try to avoid designing yourself a hazard course that you will have to negotiate on a daily basis. Avoid straight, steep hills that would be the envy of grand slalom ski racers, or running your road through low areas that will turn into huge mudholes during the spring thaw. Two important considerations: is the end of your road or driveway a safe and easy connection to the street that allows you to see and be seen by oncoming traffic in both directions? Is there a well-placed turnaround in front of your garage or parking area? (See illustrations on page 28).

The approach to your house should give you a sense of transition between the 'outside' world and your home, perhaps a gentle curve that brings you in view of your house slowly and highlights some special feature of your site, or provides a beautiful view along the way. Trees and berms can act as buffers to noise and headlights from a busy street, as well as increasing your sense of privacy. When planning where you will place your garage or parking area, try to keep it from being the focal point of the approaching

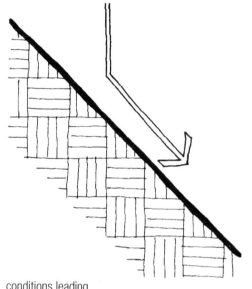

conditions leading
to increased volume and velocity of runoff and erosion: steep slopes, smooth surface, lack of vegetation, soil or substrate with poor drainage

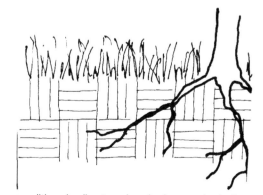

conditions leading to reduced volume and velocity of runoff and erosion: gentle or no slope, rough texture, vegetation (root mats), good drainage

Access roads and driveways can be laid out once you have defined the placement of your house. Planning in advance of any excavation or construction will ensure minimal damage to the rest of your site. Design turns and endpoints so they can be easily negotiated by the largest vehicles coming into the site during construction; this will also minimize any site destruction.

view to your house. Instead, make the front door or main entrance of your home the most important thing you see when you are approaching from the road. Too often, huge double garage doors are made the focus of approach. Ask yourself who lives here? Me or my vehicle(s)? By making the main entrance of your house the focus you emphasize the human scale of your building, which is much more inviting to visitors than an imposingly blank garage door or the bumpers of your vehicles. Creating a welcoming entry to a tight urban lot will have more challenges than larger suburban lots.

If you do not plan on having an attached garage, how are you going to make the transition from vehicle to your door with an armload or two of groceries and the evening paper in the pouring rain? Parking areas should be 15m (50 ft) or less from your door, otherwise, you are going to get very wet, unless you have a covered walkway.

Slopes, width and the radius of curves in a driveway must allow a safe entry to your garage. Driveways for garages that are located near the street on relatively level slopes require little more than the width of the car. A driveway with a slope of more than 7 per cent should be stabilized in some way to prevent erosion. Long driveways with a turnaround must be carefully designed so that you can drive in and out in a forward direction. This is much safer than backing out into a street or roadway. For safety, steep driveways should have a level area 3.6–8m (12–16ft) long in front of the garage. Garages and parking spaces for wheelchair users and other disabled people must be wider than normal to allow easy access.

SITE MODIFICATION AND REPAIR
'We must treat every new act of building as an opportunity to mend some rent in the cloth. Each act of building gives us the chance to make one of the less healthy and ugliest parts of the environment more healthy...'
From: 'A Pattern Language' (Alexander et al)

SITE MODIFICATION AND REPAIR

When you say: 'this spot is beautiful, it has everything I want – a view, beautiful trees, lots of wildflowers … this is where I'm going to build!', stop for a moment and think about what will happen if you do build on that spot. First, some or most of your lovely trees may be lost to your house or your septic field. You will lose many small plants, like that tiny patch of ladyslippers under that stand of softwoods. The view will stay, but it may not be as nice when the natural 'frame' of trees and wildflowers is gone. Animals and birds that inhabited the area will move on. In short, if you build on it, the most beautiful spot on your site may lose those qualities that you like so much.

Look around your site for areas in need of 'repair.' Imagine developing those parts of your site – building your house in that old stand of pines that is starting to die, building a rock wall to retain an eroding slope, planting a windbreak of trees and shrubs to protect your garden area. You can retain, and enhance, those aspects of your site that you enjoy the most, while slowly modifying and repairing the rest of the land you live on – whether it's a tiny city plot or several acres of varied terrain.

Repairing and modifying your site requires long-range planning – what will your changes look like in ten years? Twenty? Take your time: you can't rush the growth of a tree, or anything else that lives with the rhythms of the seasons. On the other hand, you can destroy a long-established ecosystem in a matter of seconds with a bulldozer or excavator. Plan carefully and slowly – once you've cut a tree or filled in a bog, you can't undo the change!

erosion

Because erosion can undermine the foundation of your house, cause flooding, make roads and driveways dangerous, and turn your garden into a dustbowl, repairing damage caused by erosion is a crucial form of site repair and modification.

Erosion is the loss of topsoil from an area resulting from a combination of soil conditions, slope and exposure to wind and/or water action. Insufficient groundcover, caused by deforestation, soil compaction, overgrazing or drought cycles, as well as poorly designed roads or earthworks can begin the erosion cycle. The groundcover becomes more sparse as the topsoil and nutrients are washed or blown away, exposing the less permeable subsoil, which cannot absorb water as well as the topsoil. This results in more runoff and less water retention, which causes more soil to wash or blow away, and the erosion gets worse.

The first step in controlling erosion is to attempt to stabilize the soil itself by bulking it up with organic matter such as compost. Organic matter strengthens soil by improving its structure and by increasing its infiltration properties. As well, there are species of plants that thrive in almost any type of soil – usually we class them as weeds and destroy them. But if you have erosion problems, letting weeds (or wildflowers or native grasses) colonize the eroded area can begin the process of soil stabilization by providing a fast-growing groundcover whose root systems literally tie down your topsoil.

Photo courtesy of Abri Sustainable Design.

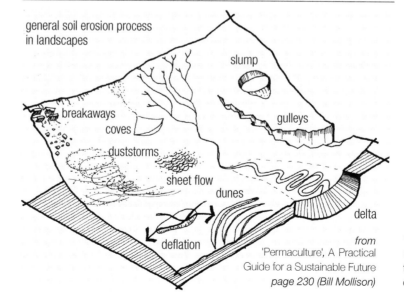

general soil erosion process in landscapes

from 'Permaculture', A Practical Guide for a Sustainable Future *page 230 (Bill Mollison)*

You can control wind erosion by planting windbreaks or fast-growing trees, or fast-spreading native grasses that stabilize the topsoil with their 'rootmats'. To control water erosion, plantings and site modifications such as terracing or diversion swales can be employed.

earthworks

You can use earthworks to improve your site in several ways:

- to reduce the energy required by your house through berming north walls against winter winds
- to repair erosion damage and control drainage through contour banks, ditches, terraces and swales
- to level out a portion of your site for placement of your house or road access
- to stop noise pollution with embankments

The topsoil of newly constructed earthworks is extremely vulnerable. To stabilize it, plant fast-growing seeds or seedlings as soon as possible after completion. Quick reseeding of earthworks accomplishes two things: it prevents erosion, which can be severe on bare slopes at only 2 per cent slope, especially in rains; and it prevents weeds from colonizing the bare soil.

PLANNING EARTHWORKS

Plan your earthworks well in advance of the arrival of bulldozers, excavators and dumptrucks, or your unwitting friends who have offered to help you 'do a bit of digging'.

Decide what you want, and roughly lay its position on your site plan.

Test the soil, or have it tested professionally before deciding on the final position (unless you are extremely familiar with soils, we recommend that major earthworks be tested professionally for their bearing capacities).

Stake and flag your site, making sure that the flagging is extremely visible for the machine operators.

Plan where you are going to store the topsoil which will be removed during excavations (for earthworks or house foundation). Never allow topsoil to be mixed with subsoil. Instead, carefully remove it so that you can put it back into your site as a growing medium and to stabilize subsoil erosion. A little extra time spent here will save you money when it comes to backfilling or landscaping.

See Appendix G for more about earthworks.

Earthworks and retaining walls are a part of nearly every building project. When building a solar house, they become an important part of the energy conserving aspect of the house. The placement and construction of any earthwork or retaining wall should enhance the 'suntrapping' capacity of your house while also protecting the house from winter winds.

soil class*	dry soil	moist soil
sandy gravel	Loose stones and single grains which feel gritty. Squeezed in the hand, the soil mass falls apart when pressure is released.	Forms a cast which crumbles when touched. Does not form a ribbon between thumb and forefinger.
silty sand	Aggregates easily crushed; very faint velvety feeling initially, but with continued rubbing, the gritty feeling of sand soon dominates.	Forms a cast which bears careful handling without breaking. Does not form a ribbon.
sandy silt	Aggregates crushed under moderate pressure; clods can be quite firm. When pulverized, soil has velvety feel that becomes gritty with continued rubbing. Casts bear careful handling.	Cast can be handled gently without breaking. Very slight tendency to ribbon. Rubbed surface is rough.
clayey silt	Aggregates are firm but may be crushed under moderate pressure. Clods are firm to hard. Smooth, flour-like feel dominates when soil is pulverized.	Cast can be handled without breaking. Slight tendency to ribbon. Rubbed surface has a broken or rippled appearance.
silty clay	Aggregates are very firm; hard clods strongly resist crushing by hand. When pulverized, the soil takes on a somewhat gritty feeling due to the harshness of very small aggregates.	Cast can be handled freely without breaking. Will form a ribbon with a slightly gritty surface when dampened and rubbed. Soil is plastic, sticky and puddles easily.
clay	Aggregates are hard; extremely hard clods that strongly reisist crushing by hand. When pulverized, it has a grit-like texture due to the harshness of very small aggregates.	Casts can bear considerable handling without breaking. Forms a flexible ribbon which retains its' plasticity when elongated. Rubbed surface has a very smooth, satin feeling. Sticky when wet and easily puddled.

*Soil particles are identified by grain size:

GRAVEL: particles over 2mm in diameter

SAND: finest particles visible to the naked eye; 0.05 to 2mm in diameter

SILT: invisible to the eye, but can be felt when rubbed in hands; 0.002 to 0.05mm in diameter

CLAY: smooth & floury or clumpy when dry; plastic and sticky when wet; 0.002mm & smaller

visual tools

Maps can aid in the site designing process. If you have an accurate survey of your site already, you can easily add your own information to it. Geological surveys usually show contours and features such as streams or riverbeds and lakes, as well as legal boundaries and setbacks. Urban and suburban plot plans or site plans often don't show any distinguishing markings and may or may not have contour lines.

To read the survey, you need to be able to pinpoint some features that allow you to orient yourself and your survey to the site. Look for the corner markers of your site. Surveyors leave pins at the corners of each site. If your site was measured a long time ago, these pins may be difficult to find or may have disappeared altogether. You may have to have the site resurveyed, as otherwise you will be unable to define setbacks, covenants or rights-of-way that are indicated on the existing survey.

Geological surveys and topographical maps show contour lines: continuous imaginary lines that connect all points of equal elevation on a site. They are set at specific intervals of 'rise', and they run at right angles to the slope of the ground. Steep ground is indicated when the lines lie close together; parallel lines indicate a smooth and regular surface; flowing lines indicate rolling hills and wriggly lines indicate broken ground.

Reading surveys accurately takes time and experience, use them to augment your on-site research to avoid any mistakes. You can easily misread the direction of the slopes on paper, while they are obvious on site. Also, maps don't show views and other features.

Use your survey or site plan as a base to indicate slope rations, rock outcroppings, drainage patterns, tree species and heights, soil conditions, special views, microclimates and noise sources and other features. Overlay your survey with sheets of tracing paper and block out the various features with markers, coloured pencils or crayons. When the overlays are all set on the survey, you will see the areas appropriate for your house, your disposal field, your well and your access road or driveway.

Photographs and videos are great aids in the site designing process. Make notes about every photograph you take, and keep up a running commentary in your site video. Talk about the character of the site, unique locations, special views, problems and potential sites and paths. Quick sketches also make you observe the site more carefully and help you to fix the features more clearly in your mind.

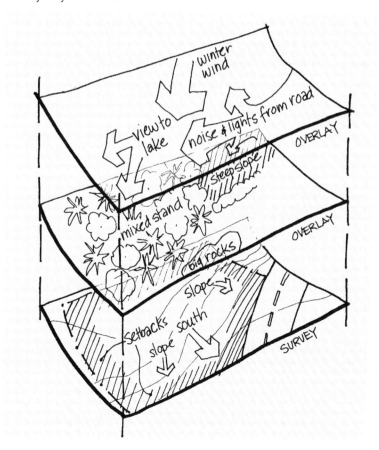

VISUAL TOOLS
Photographs and videos are great aids in the site design process. Make notes about every photograph you take, and keep up a running commentary in your site video. Note such aspects as the viewpoint (looking from the south to the dry streambed, looking east across the pile of big granite boulders beside the huckleberry bushes, etc.), the character of the site, unique locations, special views, problems, and potential sites and paths. Quick field sketches also make you observe the site more carefully, and help you to fix the features more clearly in your mind.

SITE PLANNING CHECKLIST

LEGALITIES

Note surveyed boundaries, easements and other rights, zoning and other regulations that influence site use and character, economic value. Understand all aspects of local zoning bylaws that affect your site.

- *Is your survey correct?*
- *Have you contacted the appropriate authorities about your on-site disposal system and well locations?*
- *Do you know where the municipal water and sewer mains come onto your site?*

GENERAL SITE CONTEXT

CLIMATE

Note the regional pattern of temperature, precipitation, sun angles, cloudiness, wind direction and speeds, snowfall.

Note the local microclimates: warm and cool slopes, wind deflection, local breeze, shade, plant indicators, snow drifting patterns, air quality, noise sources and sound levels.

- *Is there adequate solar access?*
- *Is there a south or southeast facing slope that is suitable to build on?*
- *How much natural protection from winter winds does the site have?*

VEGETATION

Note the dominant plant and animal communities (location, range, stability, growth patterns, regeneration potential)

- *Will important plant and animal communities be disrupted?*
- *Will it be difficult for them to relocate or to regenerate?*
- *Will rare or endangered species be destroyed?*
- *Is there adequate and appropriate soil to plant new or different species?*

WATER SOURCES

Note open bodies of water, springs, neighbouring water table levels and seasonal fluctuations, purity of water sources.

- *Will building affect the qualities of surface waters?*
- *Do streams or rivers flood?*
- *Are there any visible signs of erosion around bodies of open water?*
- *Will groundwater be contaminated?*
- *Will recharge areas for aquifers be affected?*
- *Will fluctuations in water table level affect vegetation, basements or foundations?*

There are many excellent books and videos available on landscape and garden design. Online resources are growing as well. Use the headings in this list, as well as 'site planning' and 'landscape design' as keywords for searches.

DRAINAGE PATTERNS

Note the amount and direction of major drainage patterns, any blockages, flood zones or undrained depressions (boggy areas)

- *Can the drainage patterns handle additional runoff from roofs of buildings?*
- *Can drainage patterns be modified?*
- *Can drainage patterns be used to charge a marginal well?*
- *Is runoff contaminated?*

SOIL/GEOLOGY

Note the soil types and depths, the absorption and percolation rates of different areas and stability of the soil.

Note any rock formations, as building on these will increase expense and complexity of foundation excavations, disposal field layouts and well drilling.

- *Are there any polluting sources nearby?*
- *Are there high levels of hazardous chemicals or contaminants in the soil?*
- *Can the soil handle the load of a disposal field?*
- *Are there any visible signs of erosion?*

TOPOGRAPHY

Note the pattern of landforms, slopes and orientation, any special features to be preserved, views, good access points any barriers.

ACCESS

- *Will present or planned road and utilities serve the site without adverse impacts on adjacent areas?*
- *Will it be expensive to build roads and to bring utilities onto the site?*

SUMMARY

Identify key points: solar access; orientations; views; areas best left undeveloped; areas in need of site repair; the historical context of the site and the surrounding area.

Make a summary of the positive and negative impacts of developing your site. What can you change? What can you live with? What do you want to keep? Designing is not a linear, step-by-step process – you may have to go back and rethink parts of your site design several times before it all comes together.

See APPENDIX F – RURAL SITE CONSIDERATIONS for information about the water cycle, water conservation, soils and on-site waste disposal

HOUSE DESIGN & PLANNING

In the beginning, there was the building site, and, of course, you. The site design section took you through the process of understanding and adapting you and your ideas to nature's reality. House design extends this process. Your home should make the most of your site's strengths and compensate for its weaknesses. So how does your site affect the design of your home?

Solar access will create potential heat and light for your home. For passive solar heat, you want to maximize south-facing glass. During the summer, you want to limit the amount of direct solar gain into the house and encourage cooling cross-ventilation. For light, why you need it – and when – will determine glass placement. To optimize solar hot water and PV electricity production, you want an appropriately sized south-facing roof area at a slope roughly equal to the latitude of your region.

There are three main 'streams' of solar technology for buildings: passive solar design, solar thermal systems and photovoltaic power generation. All three of these streams have design considerations that need to be included in planning your new home or renovation.

windows and solar gain

Surprisingly, 'daylighting' your house is not dependent on the size of your windows but is, instead, determined by what you do with the light after it enters the building. Window size, placement and glazing for passive solar heating or cooling and daylighting might be at odds with each other. Just to keep you on your toes, there is another consideration for your windows: view.

Windows are expensive, and, for the most part, they are also the weakest part of your thermal envelope. So, to be efficient, we want to get the highest number of functions (solar heat gain, ventilation, day lighting and view) out of each unit. We need to know the site's solar potential to be able to properly balance the sizing, placement and specifications of windows.

Site design makes you aware of where the sun rises and sets on your site and how that changes with the seasons. You should have a sense of where the sun will be at any particular time of day at any season. An awareness of obstructions to sunlight is also key to window placement. Which trees or buildings block the sun throughout the year?

In your passive solar home, you want to maximize light and energy gain in winter and minimize gain and overheating in summer. For solar thermal applications (solar air or water heating) and PV systems, you want to maximize the annual contribution of the systems. In all three applications, the determining factors are: how long and how directly the available sun strikes the window or collector, and whether the window or collector is vertical or sloped.

> Horizontal shading (i.e., extended eaves or trellises) is at its full extent at midday of the summer solstice. It offers little shading during the morning, or, more importantly, during the mid to late afternoon. Horizontal shading is even less effective in mid-August when the sun is lower in the sky and the temperature cycle peaks.
>
> Using the gable end of a house to offer shading to south-facing windows creates shaded areas throughout the day, and depending on the extent of the overhangs, throughout the hottest part of the year.

For more detailed information on windows, please turn to the building envelope section.

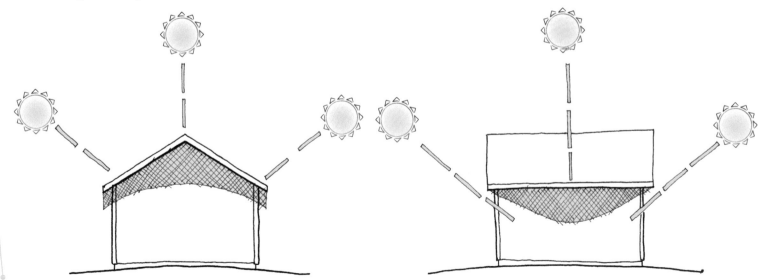

vertical glazing

The closer to perpendicular the sun's angle is when it strikes the glazing, the more energy will pass through. To capture the maximum energy and light from the low winter sun, the glass should be between vertical and 60° from horizontal. Sloping the glass more than this will decrease winter gain and increase summer overheating. The most common design error in a solar home is setting the glass too close to horizontal. Even 45° sloped glass reduces winter gain and makes it difficult to control summer overheating.

Vertical south-facing glass minimizes summer overheating. It captures most of the reflected radiation but gathers little diffused sky radiation. Vertically installed windows are also standard construction practice – no tricky building details.

	Percentage of glazing in South-facing floor area	Solar contribution
Conventional design & window placement	4 - 8%	<10%
Conventional design with better window placement	up to 15%	25%
Passive solar design with air recirculation & heat storage	15%	50%

PASSIVE SOLAR POTENTIAL
With good insulation levels and air change rates, any given house with decent solar orientation can have reduced heating costs.

sloped glazing

Sloped glass increases summer overheating, captures little reflected radiation, and gathers more diffused solar radiation. Sloped glass requires specialized construction to prevent leakage and must be installed in a vented frame assembly to keep sealed glass units from overheating and losing their seals.

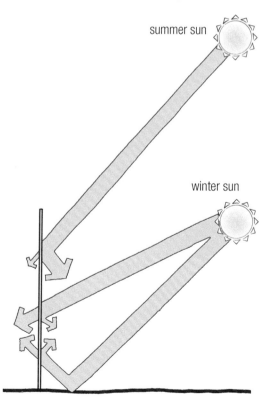

summer sun

winter sun

minimal direct gain during summer

maximum direct and reflected gain during winter

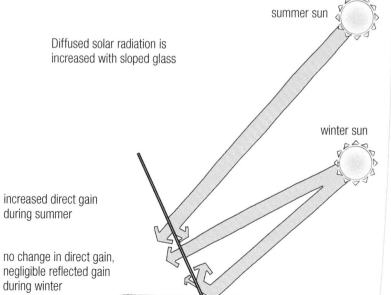

Diffused solar radiation is increased with sloped glass

summer sun

winter sun

increased direct gain during summer

no change in direct gain, negligible reflected gain during winter

In winter, windows capture the most solar gain if they face the sun when it is at its strongest. This occurs at midday when the sun is true south and its light travels the shortest distance through the atmosphere.

Site Design discusses methods for finding the strong south sun.

solar orientation

PASSIVE SOLAR

While true south is ideal, facing 15° east or west of south will have little effect on the amount of solar gain in your home. If you are more than 30° east or west of south, winter heat loss will exceed heat gain when using 'south facing' double glazed windows.

Orientation to south affects the time of day when peak solar gain occurs. For every 15° off south, the time of peak gain changes by one hour. When facing 30° west, peak gain occures at about 2p.m. In summer, this means getting maximum solar gain during the hottest part of the day, causing your house to overheat. If the windows face 30° east of south, peak gain occurs at about 10:00 in the morning, when the air is cooler. If your southwest and northwest walls have minimal glazing, or if the glazing is well-shaded, you can prevent the late afternoon sun from causing overheating.

The further off south you orient your main heat-collecting glass, the lower your overall winter gain, while the higher your summer gain. You can see that the orientation of a solar home's main windows affects both the amount of solar gain and the time of day it occurs.

Remember, the sun travels 15° across the sky in one hour.

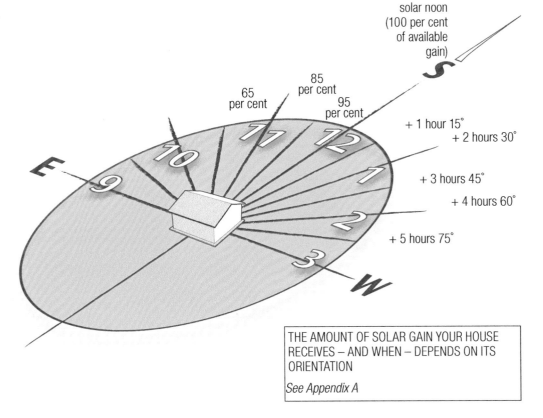

solar noon (100 per cent of available gain)

85 per cent
65 per cent
95 per cent

+ 1 hour 15°
+ 2 hours 30°
+ 3 hours 45°
+ 4 hours 60°
+ 5 hours 75°

THE AMOUNT OF SOLAR GAIN YOUR HOUSE RECEIVES – AND WHEN – DEPENDS ON ITS ORIENTATION

See Appendix A

SOLAR THERMAL

Like passive solar orientation, facing true south gives you the optimal potential solar gain for solar thermal applications. Hot water systems are typically mounted on the roof of a home, and need a minimum 3 by 3m (10 by 10ft) area clear of obstructions and penetrations. These factors don't always happen with existing homes, nor can they always be incorporated into new construction. Facing east or west reduces the potential for solar thermal systems, but, unlike passive solar gain, west is best in this case. Putting a solar thermal system on a west-facing roof allows you to take advantage of the afternoon heat. If you are mounting PV or solar thermal collectors on your roof, then the slope of the roof is a key element: the pitch should be close to the latitude of the location. This allows you to mount the collectors directly to the roof, eliminating the need for racking systems. Collectors mounted on racks can add to wind loading on the roof and foster other complications.

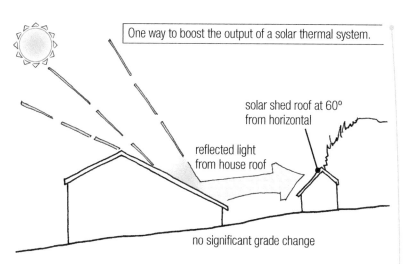

One way to boost the output of a solar thermal system.

solar shed roof at 60° from horizontal

reflected light from house roof

no significant grade change

PHOTOVOLTAICS

PV cells utilize light, not heat, so avoiding shade is the key to optimizing these systems. South-facing roof areas free of penetrations are again a good place to mount these systems, as typically, the roof of a building is slightly higher than most plantings and out of the way of most glass-breaking activities.

Photo courtesy of SHIP Database.

Photo courtesy of Blue Moon Enterprises.

The range for orientation of a solar house is between 11 o'clock and 1 o'clock, when 12 o'clock represents true south. This is equal to 30° either side of south on a compass.

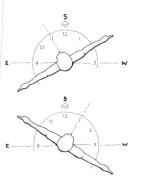

orientation and views

Orientation and view do not necessarily coincide. While the sun must squarely strike your windows for the best solar gain, you can look out the window at any angle to see the view and create views at almost any angle by stepping the profile of the house. This way, you can adjust your viewing angle to optimize both solar orientation and view. You can also rotate your orientation, losing some solar gain, in favour of a better view. As your views become further off south, it becomes more difficult to get passive solar gain. North views get no solar gain, but high-performance windows can significantly reduce the amount of heat loss on this side of the house. Houses with north views and little interest to the south can reap the benefits of solar thermal and PV systems.

A view doesn't have to be a sweeping vista. Small, intimate 'zen' views can be featured through windows in transition areas such as stairwells and entry halls. Think of windows as framing brief snapshots that are seen only when moving through a space.

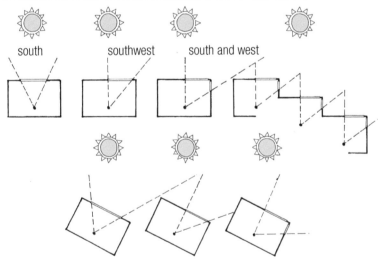

Photo courtesy of Conserve Nova Scotia.

Shift the glass along the south face of the house to access views while maintaining maximum solar gain. Where this creates semi-enclosed courtyards, incorporate shading devices on the south and west facing windows to prevent the seals on multipane glazing from 'cooking'.

south   southwest   south and west

Shift the glass and rotate the orientation of the house to increase access to views. This compromises your solar gain and changes the time of day you get it.

orientation and slopes

If you know how steep the slope is on your site, you should also know the orientation. A steep north-facing slope will have limited – or no – solar access. East and west oriented slopes may require particular patterns and geometries to optimize glazing orientation for solar gain and views. In some cases, passive solar opportunities will be minimal at best, so working with rooflines to optimize the contribution that a solar thermal and/or PV system can make might be the only way to harvest the energy from the sun.

This house is built into a north-facing slope. Views are to the north, but passive solar is optimized by a sunlit stairwell on the south side of the house that spills light and solar gain into the main living area as well as into the loft, as shown in the cross-section drawing on this page.
Photos courtesy of Abri Sustainable Design.

A steep north-facing slope will block the low winter sun

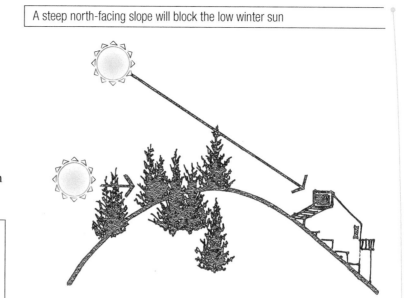

NORTH FACING SLOPE.
Light through open stairwell and clear storey in loft

summer sun

winter sun

summer sun

winter sun

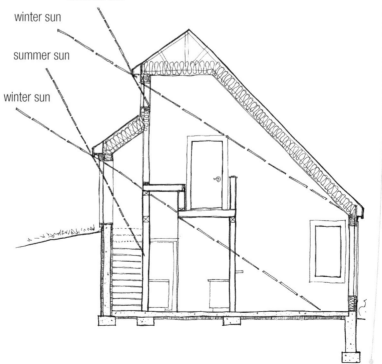

In order to minimize north-facing glazing I used single casements on the north side of my house. If I had used sliders or single-hung windows I would have required double the window area in bedrooms in order to meet building code egress requirements.

Ralph Doncaster

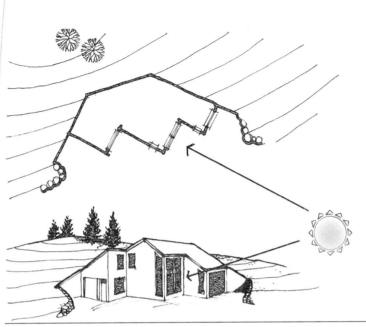

East and west facing slopes can make the design of a solar house more complex

Photos courtesy of Conserve Nova Scotia.

Angled façades and partially enclosed spaces off living rooms or dining rooms can provide two sides to a screened porch or sunspace area, giving a transition space between indoors and out. Seasonal shading devices such as a shade cloth are easily attached, and help eliminate summer overheating.

The façade of this house is angled to catch the sun's energy, and the house is also bermed into the ground to the north-facing sides, minimizing the wall areas exposed to extreme air temperatures.

obstructions

The amount of sunlight your home will receive may be limited by obstructions such as deciduous and evergreen trees, buildings, and the terrain surrounding your site. Knowing true south, you can determine if these obstructions will shade your home and roof during the critical two hours either side of midday, or 30° either side of true south by using the 'half-height rule' (page 20).

The half-height rule (page 20) does not take into account the slope on your site. If you have determined the slope and can roughly determine the height of the obstructions, AND if you are up to a little geometry, you can map out a more precise version of the half-height rule. Draw a scaled side elevation of the tree or other obstruction, place the house at the appropriate distance away from, and above or below, the obstruction as your slope dictates. Then, lay in a line roughly equivalent to the sun height at noon on January 21st. This should touch the very top of the obstruction and angle down towards the house. If this line touches the house, you will experience shading.

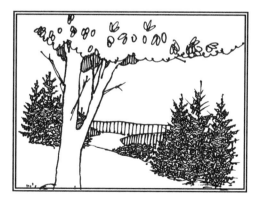

Views can range significantly from wide to tall and from near to far. It depends on how you frame them. You need to break the potential views into three elements: the foreground, which is the area immediately in front of your window; the middle ground, which is the area between the foreground and the horizon line; and the background which usually defines the horizon.

41

views

You might want to create an intimate relationship with your site by removing everything but the foreground. The resulting near and limited focuses will tie you to the ground, put you in the trees or amongst the rocks or on the water. If your view is very intimate, you will be able to see a spider on a blade of grass, or the petal of a flower. You can move that intimate focus further away from you, or you can add a series of receding focuses that will enhance your sense of distance and depth. The middle ground offers a variety of focal points, which will make your view more interesting.

If you frame your view to remove the foreground and the middle ground, you remove the visual reference points that define distance. This removes your immediate connection to the ground, leaving you 'flying' over the view. This can be exciting if you have a sweeping, panoramic view.

You can have a space that has very different views in different directions or you can have many spaces share the same view. When mixing views, choose one view as a prominent focal point. Use the remaining views in a way that creates flow and balance. You might go for diversity and give every space a different view. In all cases, you will be designing

with these three considerations: the size and distance of the view you wish to frame; the size and shape of the frame opening; and the viewer's position relative to both the view and the opening.

Photo courtesy: Abri Sustainable Design.

Photo courtesy of SHIP Database.

Photo courtesy of Conserve Nova Scotia.

Use your camera to frame views – go big, go small, shoot it all...

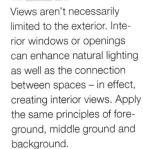

Views aren't necessarily limited to the exterior. Interior windows or openings can enhance natural lighting as well as the connection between spaces – in effect, creating interior views. Apply the same principles of foreground, middle ground and background.

slope

Very few sites are absolutely flat and level. Most have some slope. Site design shows you how to measure these slopes in 'rise to run' ratios. These ratios have many implications for your home design. To prevent flooding, water and ice damage, the ground surface around your home must carry surface water runoff away from the building.

Drainage becomes a greater problem uphill of a building as the slope of the hill increases. To begin with, steeper slopes have increased surface runoff. On a steep hill, creating a slope-away (see next page for definition) cuts into the hill. A retaining wall is then required to stabilize the cut. The slope-away cut and the retaining wall compound existing drainage requirements.

A system must be designed that will drain the resulting surface runoff from the uphill slope, the slope-away cut and the retaining wall. It's not enough just to drain the water away from the building, it must also be drained in a way that prevents flooding in other areas.

On steep sloping sites, it is best to have an experienced excavator operator and an engineer check your situation, as there are several other factors to consider, including the stability of the soil and its bearing capacity.

To prevent undo settling, the foundation of your house must bear on solid, undisturbed or compacted ground. If this bearing is close to the surface of a steep slope, there is risk of the ground being pressured out the side of the slope. On slopes greater than 1:50 (2 per cent), the foundation must bear on ground that is away from, and into, the slope. This is called a bearing setback. For safety and stability, consult an engineer when building on a sloped site.

See Appendix G for more information about slopes and earthworks.

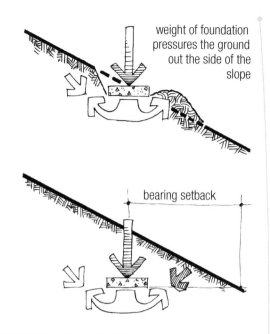

weight of foundation pressures the ground out the side of the slope

bearing setback

THE SLIPPERY SLOPE TO HIGH COSTS

When you account for slope-away, bearing setback, access to the house and outdoor spaces, designing with slopes can become very complex and restrictive. The slope does not have to be very steep before it begins to limit floor and house dimensions. With steep slopes, you may have to narrow and stretch the design out along a contour line or go to split and multiple levels. This can add to the complexity and cost of designing and building your home.

Steep slopes can also cause problems during the construction process. You have to think the construction process through. What are the limiting factors you will come up against during the different stages of the process?

Before the excavation equipment comes on the site, you have to figure out how much material will come out of the hole, and where (or whether) you will stockpile it for backfilling later on. You also need an access route in (and out) of your site. The equipment you use must have the manoeuvrability, the reach and the size to do your job.

Once you have your foundation poured and finished, how will you backfill it? Again, the right machine has to have access to the foundation without inflicting too much damage on the rest of the site. Retaining walls and stepped foundations add their own set of construction challenges. Repairing your site after construction is also more difficult on a steeply sloped site. All of these factors must be considered when building on a steep slope and all of them can easily add to the cost of building your house.

Imposing a design – or not designing for the slope – further compounds all these challenges while adding substantial, expensive site modifications and landscaping repairs.

The 'slope away' around the perimeter of your building must be at least 1:50 (2 per cent). On flat sites, slope-away must be increased, as you have to allow for ice and snow build up in low areas. Don't forget to take into account where ploughed snow will end up.

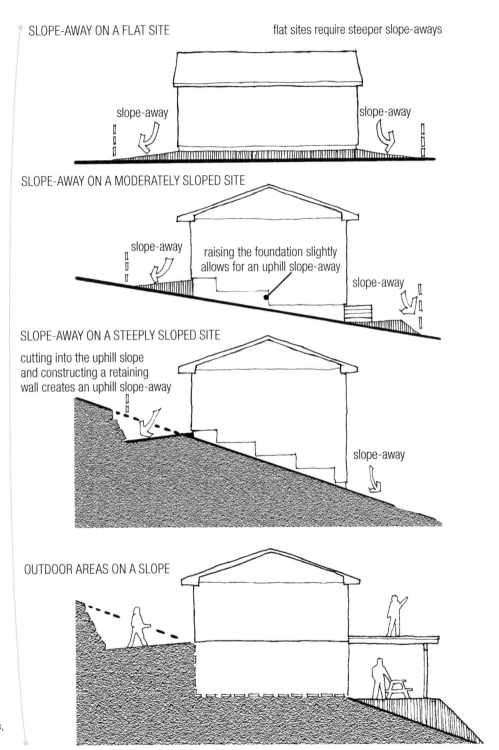

SLOPE-AWAY ON A FLAT SITE

flat sites require steeper slope-aways

slope-away

slope-away

SLOPE-AWAY ON A MODERATELY SLOPED SITE

slope-away

raising the foundation slightly allows for an uphill slope-away

slope-away

SLOPE-AWAY ON A STEEPLY SLOPED SITE

cutting into the uphill slope and constructing a retaining wall creates an uphill slope-away

slope-away

OUTDOOR AREAS ON A SLOPE

Photo courtesy of Abri Sustainable Design.

Your house design not only creates interior spaces, it also defines exterior spaces such as decks, patios and courtyards – all requiring mostly level floor areas.

Designing these on a sloped site may not only require retaining walls on uphill slopes but also aprons or elevated structures on downhill slopes. The steeper the slope the more substantial these structures will be.

Remember when you did your floor plane mockup in site design (page 21)? That exercise gives you a good sense of what the slope on your building site will require in terms of bearing setbacks, aprons and other issues.

Photo courtesy of Abri Sustainable Design.

SOME RELATIONSHIPS BETWEEN FLOOR LEVELS & SLOPES

The floors of your home will have a uniform relationship with flat ground surfaces but a quickly changing relationship with steeply sloping ground

A   there is a half-level difference between the floor level and the grade

B   there is a full-level difference between the floor level and the grade

C   there is a full level difference between the floor level and the grade. If you extend the floor out beyond the slope, the foundation must be deeper to bear on stable ground. You may also have to construct an apron to reinforce the bearing setback. This situation requires engineering

D   there is a one-and-a-half level difference between the floor level and the grade

E   there is a two storey difference between the floor level and the grade

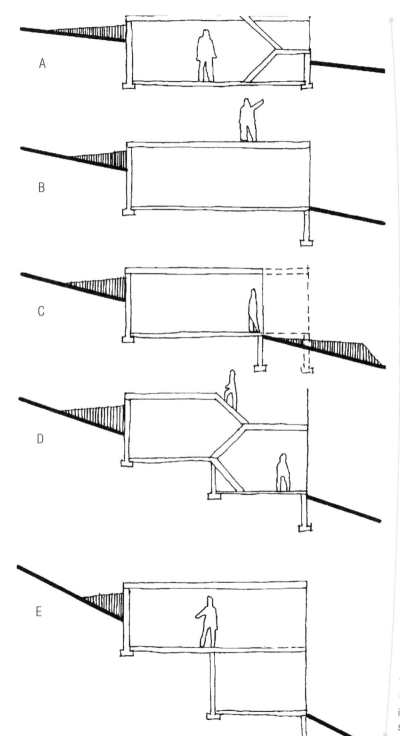

The foundation should be 'stepped' to match the slope. The steeper the slope, the more steps must be made. You can step a foundation wall down by 1.2m (4ft) increments without special structural requirements.

levels

We saw in the section on slopes how much consideration goes into setting the relationship between the ground surface and the floor level. The size and shape of the windows, and your position relative to the view, determines how you relate to your site. While your view changes depending on whether you are lying, sitting or standing the relationship between you and the view is established by the level of the floor and the window sill.

A conventional floor level is some distance above the ground. Add to this the window sill height – about a quarter of the height of the wall – and the height of the sill from the ground level

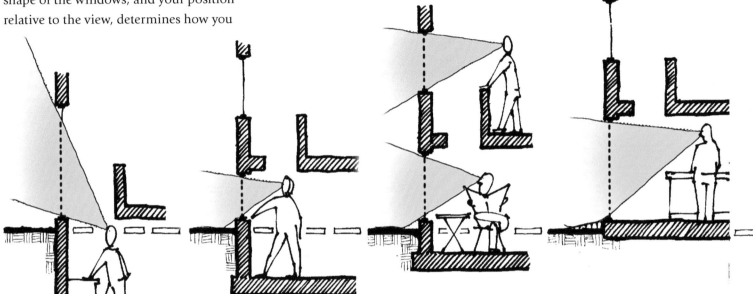

GROUND LEVEL MINUS ONE LEVEL

- no eye contact with the ground
- no view, only the tops of trees and the sky

GROUND LEVEL MINUS HALF LEVEL

- eye contact with the ground when standing only
- intimate view of surroundings

GROUND LEVEL MINUS QUARTER LEVEL

- eye contact with the ground when seated or standing
- intimate view of surroundings

GROUND LEVEL (SLAB-ON-GRADE)

- eye contact with the ground when seated or standing (through patio door)
- intimate view of surroundings
- continuity of floor level with ground level

When you drop your floor level, make sure that your window sill is not too close to the final grade, avoiding snow cover and 'splashback'. The top of the foundation must be at least 150mm (6in) above the final grade, depending on the average snow cover in your area.

becomes substantial. When you look out of this window from a seated position, your eye will meet the ground a long way out from the building. If you drop the window sill height, that distance is shortened. If your floor level is slightly below ground, and the sill is placed slightly above ground level, your eye will meet the ground very close to the edge of the house. At this distance, you can see a blade of grass, a flower or an ant on a leaf – a much more intimate relationship with your surroundings than the conventional floor level. The diagrams below explore this and other relationships between levels.

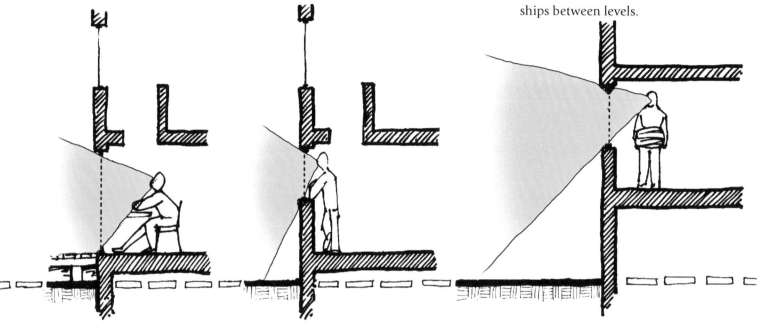

- eye contact with the ground when seated or standing (through patio door)
- intimate view of surroundings
- continuity of floor level with exterior deck, breaks connection with ground

- eye contact with the ground is distanced by window sill height
- less intimate view of surroundings, dependent on window sill height
- no continuity between floor level & ground level

- eye contact with ground is distanced by floor & window sill height
- no intimate view of ground level
- intimate view of surrounding trees, dependent on window sill height
- no continuity between floor level & ground level

Experiment with these different floor levels. Notice what rooms in different houses and other buildings feel like in relation to the ground. Also note the height of the windows and how that affects the feeling of the room – do you see more of the ground or the sky?

you and your home

While designing your home begins with your site, it ends with you.

So who are you? How will you be living in the future, how will your lifestyle change? What will your future functional and psychological needs be?

These are not easy questions. To answer them honestly, you must know yourself and your needs: how are these needs currently being met? How might they change in the future? You must understand your strengths and weaknesses in directing and fulfilling your life, and you must know how you relate to your surroundings.

Typically, you buy or rent your home. You select what is the most appropriate and affordable, and you move in. This is mostly a process of adapting yourself and your furnishings to your new home, while designing a home or renovation is the reverse: adapting the home to you. Most of us are not used to this process.

To be successful, you have to differentiate between design issues and lifestyle issues. It is very important in the design process to be realistic about how you live. For example, if your messy house bothers you, then you have to determine whether it is messy because you don't pick things up, or because there is little or inappropriate storage space in your house. Perhaps it's both. You have to be the judge of these kinds of issues.

You begin your design process with where you are. What is the existing relationship between you and where you now live? As you go about your day, think about what works and what does not work. You will become aware of hundreds of things that could be better arranged. This growing awareness and sensitivity to the built environment forms the basis of your design.

Your design should meet your household's needs, so you first need to tune into your daily routines. Become aware of the sequence of activities that happen on a regular basis. For instance, what is the weekday morning routine? How does it differ from the weekend mornings – on holidays, with the seasons, when there are guests? Will this routine change over time?

Once you have identified your routines, imagine them in detail. What time of day is it? What is the season? Where will the sun be? How might the space relate to outdoors, to other spaces? How long do you spend at various locations during your routines? Which way are you facing? Where will you go next? What is the sequence to the routine? What tools, equipment and furnishings are needed? How should they be juxtaposed to your routine and to each other?

To come up with a successful house design, you have to do your 'homework'

Observe...

Analyse...

Pretend...

Observe...

Analyse...

Pretend...

If you are more inclined to doodle on-screen, Google offers a free design tool called 'SketchUp'. It is fairly easy to learn, and allows you to draw both in 2-D (plan) and 3-D at the same time. You can create a model of your house quite quickly. As well, SketchUp files translate into most CAD programmes for final drawings.

THE DESIGN PROCESS
Of all the design tools at your disposal the most important is your creative intelligence: the ability to perceive and define a problem, to create a range of possible solutions and to isolate a successful practical approach. Every day you make hundreds of decisions, from the mundane to the profound, that affect how you live. Most of us strive to increase our ability to direct our lives and to influence our situations. In this way, we are all designers: we identify a need or a problem, we explore various possibilities, and we refine those possibilities until we come up with workable solutions. The design process is not a linear series of steps. It is a refining process that strives to eliminate undesirable or unworkable solutions. It may send you back to earlier steps to consider various ideas you overlooked or you may have to go at the problem differently. As new information is gained, new approaches or solutions may present themselves.

FLEXIBILITY, GROWTH AND OLD AGE

It is important to think of future considerations for your household. You may not always be single, a parent of small children or able-bodied. Your house may need to shrink or expand with your household. Look into ways to accommodate change in your house design. Planning for expansion is easier than shrinking but it can be done. Here are some examples of planned changes:

Roof trusses built to allow a room to be finished inside the attic space, with a large storage closet in the place of a future stairwell. This can be reversed.

Door and window rough openings for future expansion framed into the side of the house at initial construction, then insulated and finished as an exterior wall (need to be marked on the drawings so you – or future owner – don't lose track).

Unfinished basement area with walk-out door for future basement expansion.

To accommodate unforeseen changes to how the interior might be used, run engineered floor joists from outside bearing wall to outside bearing wall. Finish the interior surfaces of the outside bearing walls before installing the partition walls so that when a room size needs to be altered, you don't need to re-drywall, just repaint. This means future changes are only restricted by window and plumbing placement.

To accommodate those who cannot easily use stairs, focus on one-level living, and pay attention to door thresholds: there should be no more than 12mm (1/2in) height difference between inside and outside areas.

Wheelchair users need a 1500mm (5ft) turning radius in front of doors or work areas. This extra area reduces traffic jams and uncomfortably tight entries in general.

Curbless showers that are at least 1100 by 1500mm (3.5 by 5ft) allow a seated person to bathe unaided, and provide enough room to wash the dog as well!

For more on designing for adaptability, use the following terms as keywords for online searches:

flexible design, universal design, barrier-free, accessible.

Although there are daily routines in your life, there are also situations that are less frequent, but no less important from a design point of view. What kind of entertaining do you do? Begin to notice the dynamics of social settings. How many large or small conversation groups are there? Are people within one group aware of another group? How often do people change groups? What kind of music is playing and how loud is it? How is the lighting? And so on. When you are visiting, become aware of what you like and don't like about the space you are in and what makes it work or not work.

Design is not just observing and analysing. It is also pretending, imagining and playing. As adults, much of our imagination and many of our creative skills need to be liberated. When was the last time you played house? It is too late to find out that something does not work after you are living in your new home. By pretending, you can use the place you live now to mock up new spaces and spatial relationships. Actually sit and hold conversations at various distances – there is a range for comfortable conservation, beyond which you will begin leaning forward. Imagine various sizes of chairs, love seats and sofas, and different seating arrangements. Notice how they relate to various focal points in the room such as a window with a view or light, a fireplace, the television or other spaces. Keep your tape measure handy, to take measurements and keep notes.

Draw up furniture to scale on graph paper, make furniture cutouts, and draw up the room itself locating the windows, doors, focal points, etc. It is important to do this, as it will help you correlate real measurements with furniture and rooms drawn to scale. It will give you a sense of scale. Once you are used to relating reality to paper, you can then use paper exercises to develop the design. With a space drawn to scale and the furniture cutouts you can quickly change or rearrange furniture and test the space.

It is a good idea to use the same scale for your design drawings as the drawings from which the house will be built. Most residential drawings are done at 1/4 inch to 1 foot scale. For pattern development, use a scale of 1/8 inch to 1 foot.

Get out a measuring tape and start to find out how much room you need to move, how high is a counter, a vanity, a table, a stool, a chair, a bed, a sofa... How wide is a hall, a door, a closet, a kitchen cabinet, a particular room that you really like? By observation, you can teach yourself a tremendous amount about your design needs and about how you and others relate to each other and to spaces.

the kitchen exercise
evaluate layouts for routines, patterns & elements

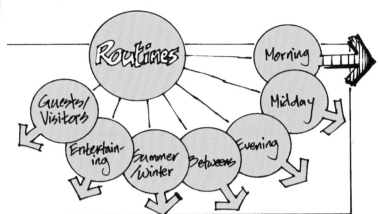

TO START, LET'S DEFINE YOUR MORNING ROUTINES...

- *who?*
- *when?*
- *how many at the same time?*
- *individual routines?*
- *collective routines?*
- *weekday/weekend?*
- *winter/summer?*
- *guests/visitors?*
- *outdoor?*

Evaluate each routine individually

The exercise on these two pages gives you an idea of the kinds of questions you have to ask yourself, and the amount of thought and evaluation you have to go through, to design a space. We have used the kitchen as the space and the morning routine as our example because everyone has a kitchen of some sort, and most people have fairly consistent morning routines.

The first thing to do is define the routines. Our example asks questions about WHAT happens in the morning in the kitchen. (See the list on the upper part of this page.)

The next thing to do is to sequence those routines. HOW do you and your household use the kitchen in the morning? What is the sequence? (See the list of activities on the upper part of the facing page.)

'L' kitchen

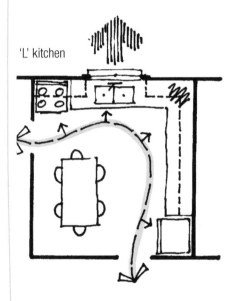

- two entrances – door access only
- window/sink room focus
- linear movement along counter
- closed to adjacent spaces
- always working with your back to the room
- kitchen table with no view
- one corner cabinet
- fridge and stove near entrances
- no dead floor space

'U' kitchen   open

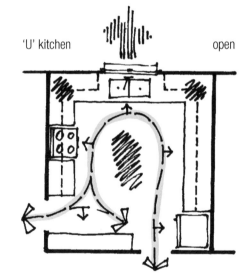

- two entrances – closed to adjacent spaces
- window/sink room focus
- linear movement with long crossing distance
- closed to adjacent spaces
- always working with your back to the room
- no kitchen table
- two corner cabinets
- fridge and pantry near entrance
- large dead space in the centre of the room

'U' kitchen   closed

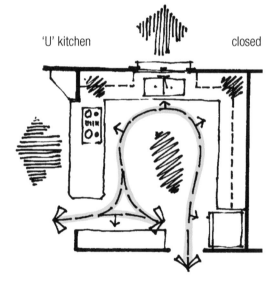

- two entrances – across counter & door access
- window/sink & adjacent space focus
- sink area visible from adjacent space
- linear movement with long crossing distance
- open to adjacent spaces
- can work facing adjacent space
- double counter for preparation/dining
- two corner cabinets
- fridge and pantry near entrance
- large dead space in the centre of the room

50

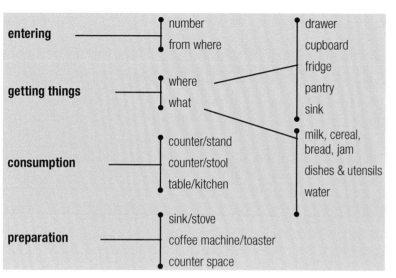

| entering | number |
| | from where |

getting things	where	drawer
	what	cupboard
		fridge
		pantry
		sink

consumption	counter/stand	milk, cereal, bread, jam
	counter/stool	dishes & utensils
	table/kitchen	water

preparation	sink/stove
	coffee machine/toaster
	counter space

Once you have defined the routines and their sequence, it's time to think about the different kitchen layouts that are available to you, and how your particular sequence of morning routines might fit into them. Think of how your current kitchen layout works – or doesn't work – with your morning routines. What could serve your needs better?

Then, it's time to think of how you use the kitchen at other times of the day, and how those routines fit into the layouts. Conflicts between different routines mean that you will have to decide which routines have more importance in your life. For instance, if breakfast is a haphazard affair in your home, but everyone sits down together for the evening meal, then you might want to design your kitchen around the needs of the evening meal.

This exercise can be used as a model for determining the routines in any part of your house. Once you know all of the routines that take place in one space, you can appropriately design it.

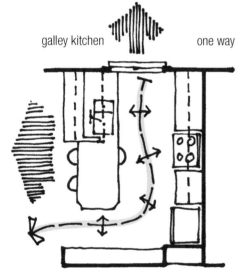

galley kitchen   one way

- one entrance – across counter & door access
- window/sink & adjacent space focus
- sink area hidden from adjacent space
- linear movement with short crossing distance
- open to adjacent space
- can work facing adjacent space
- double counter for preparation/dining
- no corner cabinets
- fridge and pantry near entrance
- no dead floor space

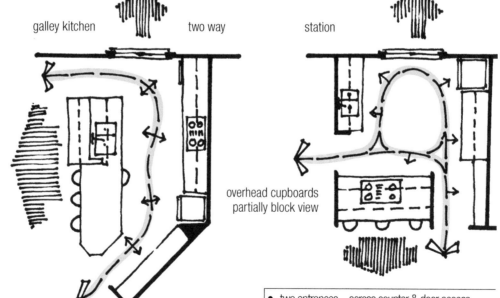

galley kitchen   two way   station

overhead cupboards partially block view

- two entrances – counter & door access
- open to two adjacent spaces
- not as efficient as one-way galley

- two entrances – across counter & door access
- window/sink & adjacent space focus
- sink area hidden from adjacent space
- crossing movement with short distances
- open to adjacent spaces
- can work facing adjacent space
- double counter for preparation/dining
- no corner cabinets
- fridge and pantry near entrance
- no dead floor space

synthesis

Most of us are used to square or rectangular rooms with four solid flat walls one or two windows and a door. So that is what we design. If we want to connect two rooms, we leave a wall out. We design with walls or no walls. But there is much between these two extremes. Rooms will be more or less separated depending on the height and length of the wall between them. With less separation a room becomes a space. While rooms are definitely spaces, spaces aren't necessarily rooms.

Well, then, what is a space? It is a defined area. Designing a space is defining its interface with adjacent spaces and with the site. This interface can be any combination of visual (light and view), acoustical, tactile (air movement and vibration), or thermal elements and cues. Transparent, semi-transparent or solid walls of all sizes, shapes, colours and textures can enclose and separate spaces. So can furnishings and plants.

The need for a space or a room comes from bringing routines together. The routines give us the functional requirements for the space and also tell us what the relationships should be between spaces. You should be aware of what you can or cannot see from adjacent spaces: light, views and focal points, or clutter,

dirty dishes and the kitty litter box. You should also be aware of what you can smell and hear.

Changing the dimensions of a wall or changing your location relative to the wall will change what you see. If you see a wall square on, your eye will just stop. If you see a wall at an angle, your eye will be directed along the wall. If you really want to get daring, instead of the mostly static effect of vertical and horizontal edges, try the dynamics of sloped edges!

Photos courtesy of SHIP Database.

Photo courtesy of Abri Sustainable Design.

enclosing spaces

A solid wall, depending on its length, width and height, will separate you from what is beyond it, and create a sort of space. Two or more walls will further separate and define a space. Putting two walls together at a slight angle will create a corner, further separating what is beyond and enclosing a space. Closing the angle will further tighten the space. Lengthening the walls will increase the sense of enclosure. Adding a third connected wall creates two corners, further restricting and enclosing the space. However, the second corner adds another element: dimension.

Dimension is the measure of length, width or height. Dimensions can be perceived visually or physically measured. The physical measure of space gives us function, while the visual perception of space gives us form.

Corners provide the beginning and end points for measuring and perceiving dimensions. The more corners, the more dimensions. The more corners, the more points your eye will perceive.

Photo courtesy of SHIP Database.

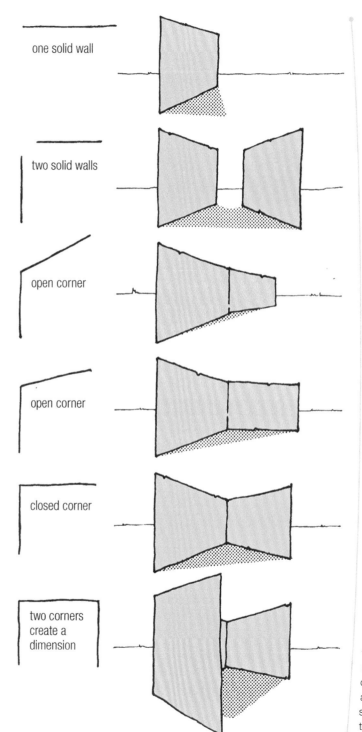

one solid wall

two solid walls

open corner

open corner

closed corner

two corners create a dimension

Corners happen wherever two walls meet, a wall and a floor meet, a wall and a ceiling meet, or a ceiling and a floor meet. A corner is formed by any angle between two planes, and is not restricted to 90° (i.e. squares or rectangles). Corners can also be rounded – no rule says the space you define has to be angular.

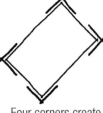

Two walls make a corner; by length-ening the wall, you enclose more space

By adding more corners, you add dimensions, which further define & enclose space

Four corners create a totally dimensioned and enclosed space – a room

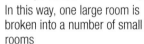

By adding more corners, you create more dimensions and more rooms

In this way, one large room is broken into a number of small rooms

A corner doesn't have to extend very far into the space to create a dimension

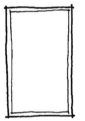
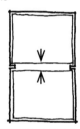

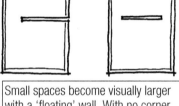

By visually adding corners (in the form of posts along a wall or beams along a ceiling), visual dimensions are added to the room. Functionally, there is a single space, but visually, there are two.

Small spaces become visually larger with a 'floating' wall. With no corner, there is no fixed point where one space ends and the other begins, so your eye flows into the adjacent space.

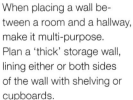

The process of perceiving a space is quite complex, but it is also something you do all the time. You may not be aware of it. Being more attentive to this perceptual process will help you create spaces, as it will tell you which elements to use and how to use them.

When placing a wall be-tween a room and a hallway, make it multi-purpose. Plan a 'thick' storage wall, lining either or both sides of the wall with shelving or cupboards.

Photos courtesy of Conserve Nova Scotia.

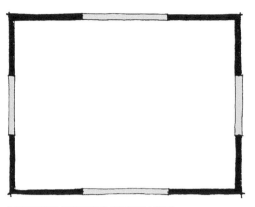

WINDOWS CENTRED ON THE WALL
- very strong corners
- very strong sense of enclosure
- visual cross-axis

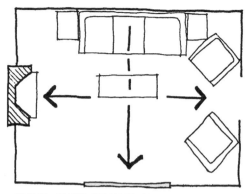

VISUAL FOCUS CENTRED ON WALL
- cross axis is shorter, room feels smaller

WINDOWS IN THE CORNERS
- weak corner
- different sense of enclosure
- visual diagonal axis

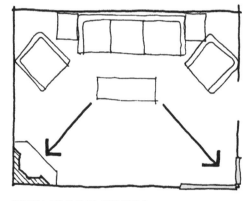

VISUAL FOCUS IN CORNERS
- diagonal axis is longer, room feels larger

If you have the opportunity to step aboard a yacht or two, check out the space planning. With both floor space and storage space at a premium, boat designers are very good at squeezing the most out of what they have to work with. If getting onto a boat is not possible, borrow a few yachting magazines from your local library.

A rule of thumb: as the complexity of the design increases, so does the cost of building your home.

55

WINDOW CENTRED IN WALL

- reflecting surfaces are distanced from window, creating a dark room

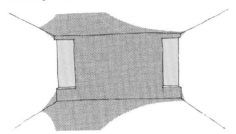

WINDOWS AT CORNERS

- room is bright when the sun shines on the reflecting walls

Photos courtesy of Rob Dumont

Placing windows close to corners allows the light to splay across a bigger wall surface. Using semi-gloss paint increases the amount of light that can be bounced into a room. A continuous surface is a better reflector than a surface broken up by framing and trim details.

groupings

Following the design of most of your spaces, it is time to begin to assemble them. Areas or rooms that are intimately tied to each other, such as an open kitchen and adjacent family room, can be treated as a single space. Spaces that are closed but functionally linked become a grouping. Thus, a closed living room, dining room and kitchen-family space can be collected as a grouping. Groupings define close links among rooms and spaces, and can be contained on one floor level. Creating groupings reduces the number of elements under consideration for your home's layout. With fewer elements, it is quicker and simpler to run through all the ways to create a successful floor plan. Any remaining unlinked spaces and groupings can often be positioned remotely from one another or put on different floor levels.

The more spaces, the more groupings and the more complex the design, the more floor area – circulation space – needed to access each space and grouping. Some spaces can open to and share the circulation space. For example, having a shared dining and circulation space allows for expansion of the dining table and chairs into the circulation space when entertaining large groups. In this way, you add flexibility while reducing the overall floor

Photo courtesy of Big Sue, LLC.

space. However, most spaces function best without sharing circulation space and without a circulation path through them. Your circulation space is best consolidated and centred into a core area.

Keeping a large number of spaces on one level will require them to be spread out, resulting in a higher proportion of circulation space. Splitting the spaces and groupings into levels may make the design more compact, but stairs make areas of the house inaccessible. Stairs affect the need for halls, which in turn affects the pattern of spaces and increases the complexity of your design. What may work well on one level may not work on another. Various stair layouts can be grouped with their halls and circulation patterns. When you add up the floor

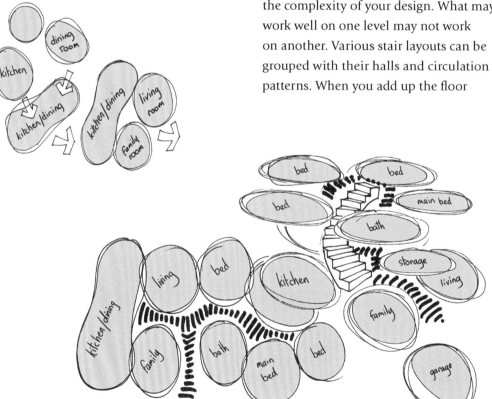

If you are building a super-insulated house with thick walls, angle the sides of the window cavities so they open wider into the room. This will increase the amount of light the window brings into the space. Straight-sided window openings, regardless of how close they are to an adjacent wall, act more like a tunnel, directing light into the middle of the room.

Choosing a stair geometry depends on where you want to be placed on each of the connecting levels.

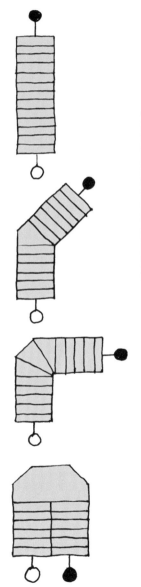

space taken up by stairs and halls, you might find you are better off leaving everything on one level.

Moving from one space to another through an opening, a hall or a stair can also be visualized as a routine. Imagine what you see first, second, third – the light, the shifting focuses – as you move from one space to another.

Stairs can be a pleasant visual transition: the opening creates a link between the various levels. This opening may also allow for free air circulation between levels. A stairwell is also an ideal location for a return air outlet – noise from a fan pushing air around will be less noticeable in such a transition space.

Photo courtesy of Abri Sustainable Design.

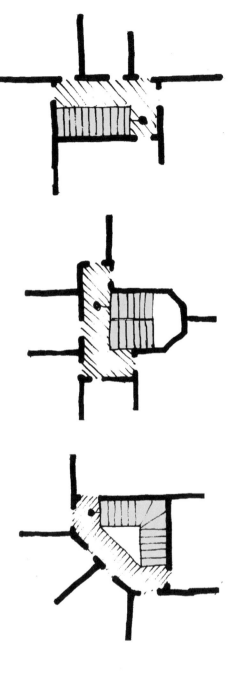

When you are assembling the pattern for your house, you must be aware of the circulation halls that are attached to the different stair geometries.

a solar pattern

After imagining the routines and designing the spaces, groupings and circulation, it is now time to bring these together into an overall pattern. Since interior interfaces can be rotated, flipped, mirrored or broken into levels, there is some flexibility in evolving a number of patterns and

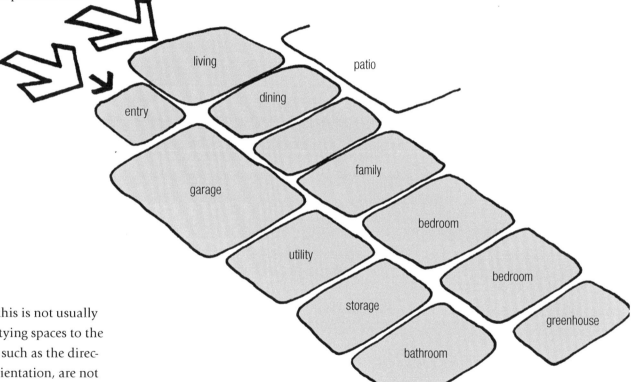

geometries. However, this is not usually the case when you are tying spaces to the site. Most site features, such as the direction of a view or the orientation, are not changeable. Others, such as the direction of access, slopes and wind direction are adaptable with great difficulty and often result in substantial site damage. The best site interface should dictate the overall pattern and geometry of the solar home. An open plan has the added benefit of better air and heat flow.

Having imagined the routines and designed the spaces, groupings and circulation, it is now time to bring these together into an overall pattern. Most site features, such as the direction of a view or the orientation, are not changeable. Others, such as the direction of access, slopes, and wind direction are adaptable with great difficulty and often result in substantial site damage. The optimum site interface should dictate the overall pattern and geometry of the house.

The sun permeates the whole design process: how most of the rooms and spaces are designed, how they are linked and interfaced, and how they are patterned.

SUMMARY OF KEY POINTS
- *Prioritize spaces for access to the sun*
- *Maximize the openness of the plan*
- *Use transparent spaces*
- *One or two level layout?*
- *Merge and optimize exterior geometry*
- *Set the house into the ground*

PRIORITIZE SPACES FOR ACCESS TO THE SUN

Not all spaces need or want direct access to the sun. Some spaces are happy to share the sun through another space. Others are happy with a view or indirect light, while a few are best left in the dark.

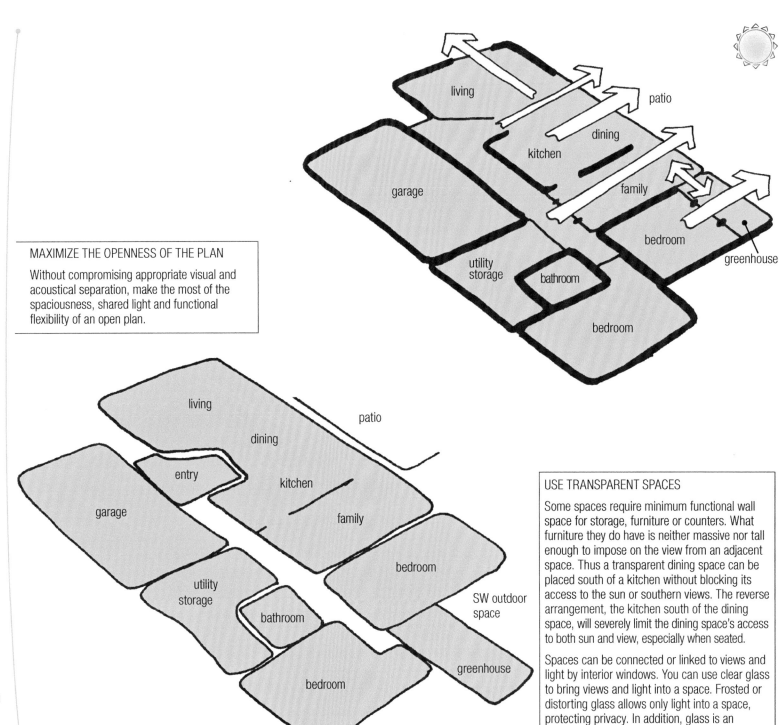

MAXIMIZE THE OPENNESS OF THE PLAN

Without compromising appropriate visual and acoustical separation, make the most of the spaciousness, shared light and functional flexibility of an open plan.

USE TRANSPARENT SPACES

Some spaces require minimum functional wall space for storage, furniture or counters. What furniture they do have is neither massive nor tall enough to impose on the view from an adjacent space. Thus a transparent dining space can be placed south of a kitchen without blocking its access to the sun or southern views. The reverse arrangement, the kitchen south of the dining space, will severely limit the dining space's access to both sun and view, especially when seated.

Spaces can be connected or linked to views and light by interior windows. You can use clear glass to bring views and light into a space. Frosted or distorting glass allows only light into a space, protecting privacy. In addition, glass is an excellent noise barrier.

An open plan allows you to optimize the floor space, making a more compact and affordable house, with the added benefit of better air and heat flow.

ONE LEVEL LAYOUTS

One level layouts rely heavily on the previous points to expose many spaces to the sun without stringing them out in a long line. While a long linear pattern may have a large southern exposure, it will also have a large northern exposure. The north exposure can be minimized by setting the building into a south slope or by earth berming. Whether you group spaces or have a linear pattern, locating utility, storage and other non-solar spaces on the north can act as a buffer. Skylights can be used to light northern spaces.

MIN/MAX EXTERIOR GEOMETRY

Your home's geometry should minimize exposure to winter winds, heavy storms and unwanted noise, while maximizing its exposure to winter sun, to cooling summer breezes, to pleasant sounds and views.

TWO LEVEL LAYOUTS

Two level layouts make a more compact envelope while exposing many spaces to the sun. However, the taller profile of the house may also have significant northern exposure. This can be minimized through the geometry of the roofline and the upper level of the house. Skylights can illuminate upper northern spaces. Use the stairwell to bring light to lower spaces from upper windows or skylights. Pulling the upper level back from the south exterior wall will allow deeper light penetration on the main level.

MERGING GEOMETRY

Your house can visually and functionally merge with your site so that it looks and feels comfortable. The geometry of your house not only creates interior spaces, but it also defines important exterior spaces. These exterior spaces can be carefully and thoughtfully designed into your home and site to create semi-enclosed outdoor spaces or 'rooms' and microclimates. House entrances, courtyards, patios and decks are important linkages between your house and your site. They integrate your home and site.

The house at the bottom of this page shows diverse relationships with the land around it. The patio doors are at floor level, the windows at the right corner and side are a foot or two above floor level, and the window at the left corner of the house is above waist height, yet all are directly linked to the landscape, informing the relationship between inside and outside.

Photo courtesy of Abri Sustainable Design.

OPTIMIZING EXTERIOR GEOMETRY

The most efficient structure in the world, in terms of possible energy use and conservation, is the geodesic dome. It holds the most volume in the least surface area possible, minimizing the proportion of the building exposed to dramatic swings in temperature. The next best shape for a building is a cube. Whatever your house shape, one of your design goals should be to minimize the exposed surface-to-volume ratio. Simple shapes work better than complex ones for many reasons: the more corners and roof penetrations you create, the higher the amount of framing required. More framing equals more opportunity for air leakage and also reduces the overall insulation value of the wall or ceiling. Slightly larger, simpler shapes will be less expensive and easier to build and maintain. Setting the house into a berm or slope reduces the exposed surface area as well. Building 'into' the attic, or creating a second storey or loft area with a sloped ceiling puts all your construction costs into living space, not attic space.

KEEP A HUMAN SCALE
A human scale makes your house feel like home. Design your house to satisfy your most important functional needs first. Then use the devices in the spatial exercises to create visual distance, interest, diversity, balance and flow without adding to the structural dimensions.

OPTIMIZE EXTERIOR GEOMETRY

Your geometric design should minimize the surface to volume ratio of the house. Reducing the amount of surface area exposed to the elements reduces the potential heat loss from a house. It should do this with a simple straightforward geometry with minimal articulation, appendages and penetrations. A slightly larger, simpler shape will be less expensive and easier to build and maintain.

SET THE HOUSE INTO THE GROUND

The deeper into the ground you go, the warmer the annual ground temperature will be, and the less that temperature will fluctuate with the seasons. In winter, a house that is all, or partially, underground will lose less heat to very cold air temperatures and stay cooler in the summer.

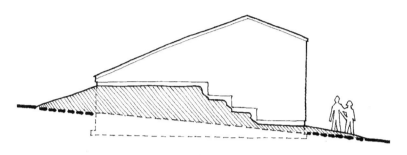

Set your house into the ground by building into slopes and by earth berming

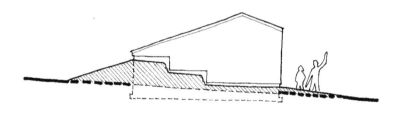

Photo courtesy of Abri Sustainable Design.

Photo courtesy of SHIP Database

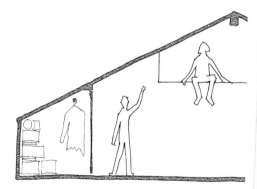

OPTIMIZE INTERIOR VOLUME

There are lots of opportunities to use the shapes and dimensions within your geometry in practical and functional ways. This is especially true under sloped ceilings. Lofts, half-height storage closets and other devices make the space usable.

Building into the attic space makes sense in terms of energy savings as well as getting the most out of your construction dollars. instead of unuseable, inaccessible attic space, you can have a bedroom, a study, a studio or an 'away' space.

planning for solar add-ons

You can incorporate 'solar add-ons' – solar thermal and photovoltaic systems – into your home design and planning process, even if you aren't installing the systems right away. Planning for these systems – making your home 'solar ready' – is a good way to anticipate future options for your energy needs, and increase the market value of your home at the same time. A small investment in plumbing and wiring during new construction or a major renovation project, means a future installation of a renewable energy system will be cost-effective and simple. Installing a solar thermal or PV system after construction can easily cost half as much as the system itself.Most often, these systems are installed on the roof of a house or an outbuilding such as a garage or a shed. They can also be installed on the ground, but you need a large area in front of the collectors that is free from shading and other obstructions. Also, putting glass-covered collectors in range of kids, balls and other projectiles might not be in your best interests. Roofs conveniently provide a clear surface at a significant height above the surrounding vegetation, most landscape features and projectiles. However, annual visual inspections and maintenance work require access by ladder.

For a rooftop system to work, the number one house design element is a roof face of adequate size and slope with decent solar access for your energy needs. There should be no obstructions from other roof elements such as dormers, satellite dishes and chimneys. The plumbing and/or wiring will need to lead from the attic or roof to the mechanical room without interfering with any structural elements such as beams, headers or posts. The mechanical room will have to be sized to include extra water storage, controls and access to both.

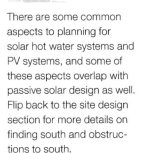

There are some common aspects to planning for solar hot water systems and PV systems, and some of these aspects overlap with passive solar design as well. Flip back to the site design section for more details on finding south and obstructions to south.

SOLAR READY

Making a home 'solar ready' is similar to roughing in a central vacuum cleaning system. Installing the central vac ducting after construction would cost a lot of money, but the rough-in package is a standard option with most builders, not even showing up as a budget line in some cases. Studies in California have shown that a solar ready house will have a higher re-sale value than similar houses in a given neighbourhood. This is certainly a factor in making decisions about the investment you make in your house.

MAJOR DESIGN ELEMENTS TO BE TAKEN INTO CONSIDERATION:
- *Siting the building on your lot*
- *The shape and mass of the roof and any living spaces it will contain*
- *The shortest path from the roof to the mechanical room*
- *The size of the mechanical room – available floor and wall space*

Services to be taken into consideration:
- *Adequate wiring for both solar thermal and PV installations*
- *Two runs of copper pipe (pressure tested and sealed if roughing in system)*
- *Install a solar by-pass at water tank if roughing in system*
- *Clearly label everything – piping and wiring at both ends of system*

Photo courtesy of SHIP Database.

top down planning

After going through some or most of the planning and design stages for your house you need to think about integrating solar thermal and PV. So, let's start at the roof. You need a clear, unobstructed roof area of at least 2.5 by 2.5m (10 by 10ft) for every two solar thermal collectors. Because of the variation in efficiency of PV modules, each kilowatt of capacity can require between 6 to 10m² (65 to 110ft²) of roof space. Additional space should be left for adding modules.

Both solar thermal and PV systems perform best closer to solar south. However, even due east or due west roofs can work for solar thermal systems, although the efficiency will be reduced. In most areas, a west-facing system can take advantage of warmer afternoon temperatures and the late afternoon sun, while an east-facing system may be limited by morning fog or mist, and cooler morning temperatures.

The collectors must be installed at an angle that optimizes solar gain for your needs. To ensure good average annual performance, your roof pitch should be within 10° of the latitude of your site. So, if you are at latitude 45°N, your roof pitch should be 35 to 55°, or 8:12 to 17:12. A suitable roof pitch eliminates the need for a racking system. Racking systems add to the loading on your roof. Whether you are installing your system now or in the future, make sure you have access to the area under the roof, with enough room and height for a person to work up there.

A standard flat plate collector is 1200mm by 2440 or 2740mm long (4ft x 8 or 9ft). A standard solar domestic hot water system will include two collectors. Evacuated tubes come in banks of 900 to 1200mm (3 to 4ft) wide by 1800 to 2440mm (6 to 8ft) long. Two banks make up a typical domestic hot water system.
PV modules are between 600 to 1200mm (2 to 4ft) wide and 1200 to 1800mm (4 to 6ft) long.

Solar thermal systems perform best when oriented within 30° of solar south, while PV systems are reasonably efficient up to 45° off solar south.

For better performance in the summer, roof slope is 15° less than your latitude. For better winter performance, go 15° higher than your latitude. North of 60°, install systems on vertical surfaces such as walls to take full advantage of the near-continual sun throughout the summer months and the very low sun angles experienced during the winter.

SOLAR THERMAL HOUSE

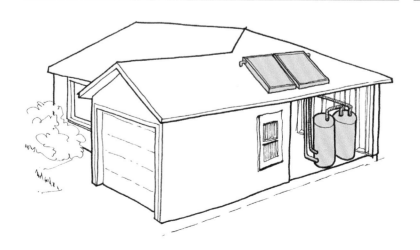

GRID CONNECTED HOUSE

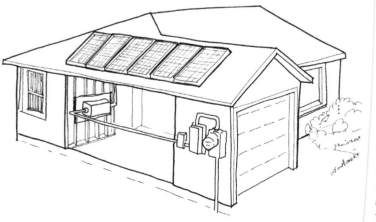

PLANNING FOR ROOFTOP SOLAR COLLECTORS

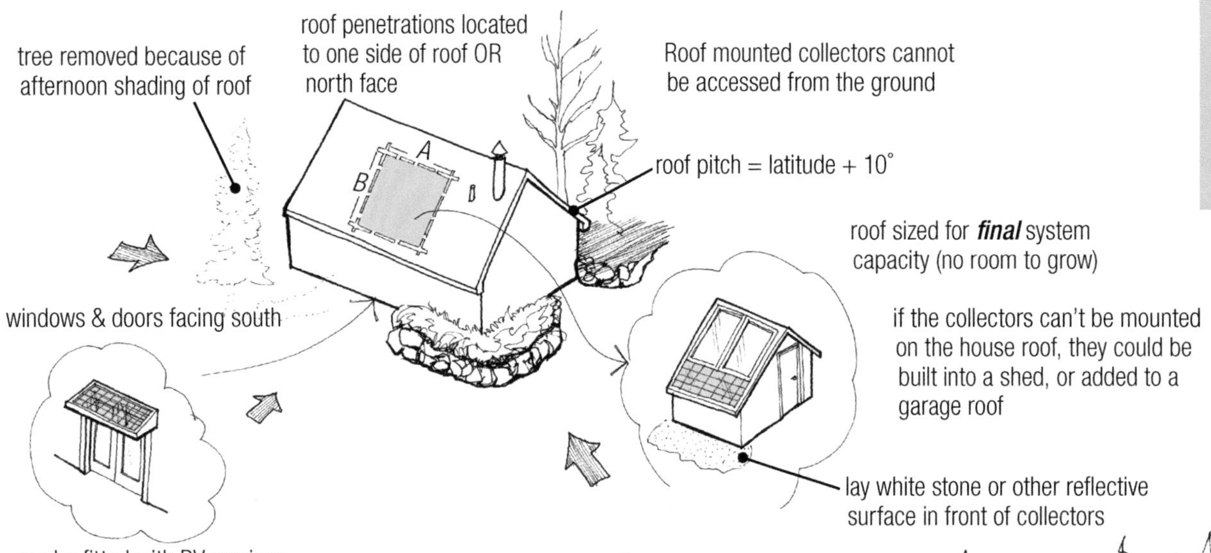

tree removed because of afternoon shading of roof
roof penetrations located to one side of roof OR north face
Roof mounted collectors cannot be accessed from the ground
A
B
roof pitch = latitude + 10°
roof sized for ***final*** system capacity (no room to grow)
windows & doors facing south
if the collectors can't be mounted on the house roof, they could be built into a shed, or added to a garage roof
lay white stone or other reflective surface in front of collectors
can be fitted with PV awnings

SOLAR COLLECTOR AREA

A = 2440–3050mm (8–10ft) for every 2FPC/ETC or for every 1kw of PV

B = 3050mm (10ft) for every FPC/ETC or for every 1kw of PV

fill
excavations
afternoon sun
private courtyard
kitchen
living room
future power pole
well
entry courtyard
new planting for visual screen
septic
field
noon sun
views
morning sun
driveway

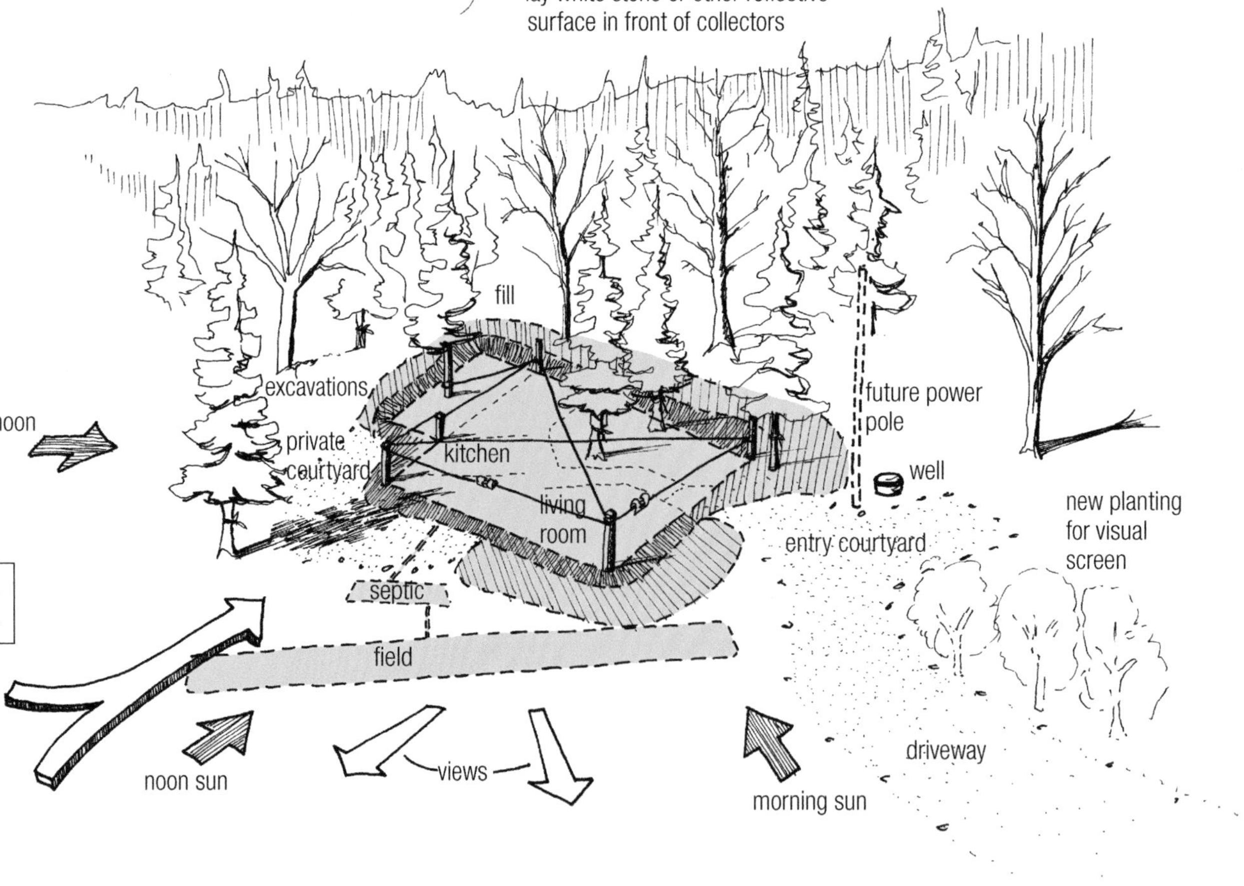

IDEAL PASSIVE SOLAR SITE IN A SUBURBAN OR RURAL SETTING.

this drawing shows the house site staked out with reference to the suns position throughout the day, the views and the prevailing winds. The floor plan mocked up with lines and string levels.

Rural and wide suburban sites have different options than narrow urban and suburban sites. However, passive solar potential can be compromised for a site because of the way that the road is oriented, or because of any covenants that might require the front face of the house to be parallel to a North-South running street.

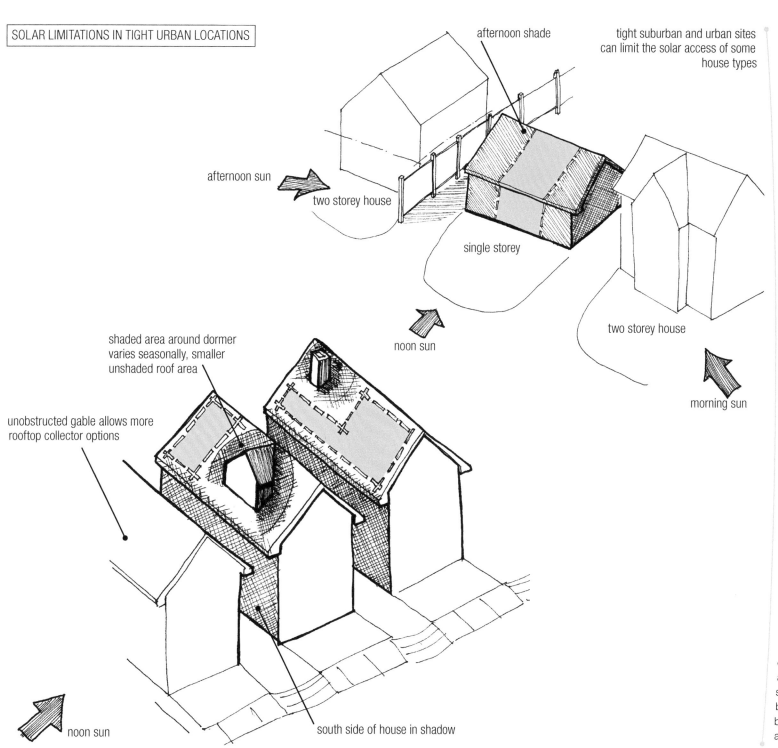

afternoon shade

tight suburban and urban sites can limit the solar access of some house types

afternoon sun

two storey house

single storey

two storey house

noon sun

morning sun

shaded area around dormer varies seasonally, smaller unshaded roof area

unobstructed gable allows more rooftop collector options

south side of house in shadow

noon sun

Houses on urban lots can have very limited passive solar access, but an unobstructed rooftop with a gable that runs East-West can be a good site for solar collectors, especially if the neighbourhood is already built up. As dormers can throw shadows across a wide area on the surrounding roof area, it is best to do a shadow study before installing collectors on a roof with a dormer.

Going down through the house, for a solar thermal system, your space planning must allow two runs of copper pipe to make it from the roof to the mechanical room. For a future PV installation, a sealed and labelled conduit allows you the most options. To get around structural members, build a chase.

Now that you know how to get from the roof to the mechanical room or utility area, it's all a matter of space planning. Solar thermal systems typically have two hot water tanks, and some have external heat exchangers. If you are planning your solar thermal system for space heating or want more storage, you will need a bigger roof area for the collectors, and more – or bigger – storage tanks.

A grid-connected PV system typically takes up no floor space, but inverter, disconnect switches and (optional) charge controller, breakers and fuses require wall space. Check with your utility about requirements for any system components that must be accessible. For off-grid or grid-intertied systems, batteries must be kept off concrete floors, and they must be housed in a secure, self contained area, with ventilation to the exterior.

In all of your mechanical room planning, remember that for maintenance and repair work, all parts of any system need to be accessible. An average adult

'U' SHAPED LAYOUT OF MECHANICAL SYSTEMS.

needs at least 1m (3ft) in any direction to be able to move from a standing to a squatting position. If you can create a 1.3 by 1.3m (5 by 5ft) area between two banks of mechanical systems, you, your plumber and electrician will be happy (and safe).

Collectors and mounting systems increase the structural load on the roof by up to 5lb/ft² (24.4kg/m²). The roof and the mounting system must also withstand uplift of the panels due to wind as well as bear snow loads. Tighter spacing of roof trusses may be enough to counteract these forces if you are building a new roof, but make sure that a qualified person is sizing your roof structure and be clear that you intend to install a rooftop system.

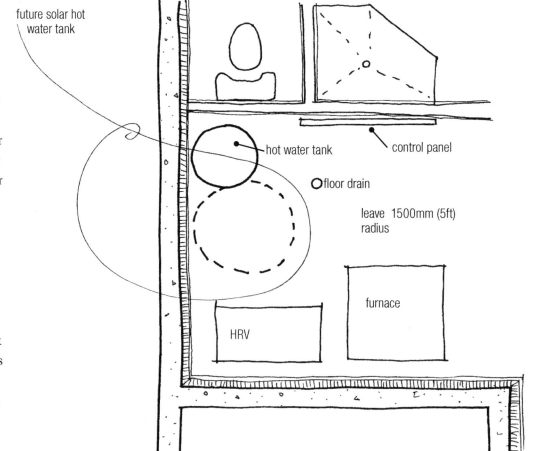

future solar hot water tank

hot water tank

control panel

floor drain

leave 1500mm (5ft) radius

furnace

HRV

SUMMARY OF DESIGN AND PLANNING POINTS

ROOF DESIGN:

Solar thermal: 3 by 3m (10 by 10ft) for every 2 flat plate collectors or bank of evacuated tubes

PV: 6 to 10m² (65 to 110ft²) for each kilowatt within 45° of solar south

Roof oriented to within 30° of solar south

Roof pitch within 10° of your latitude for best year-round performance

Roof structure must be able to hold the weight of the collectors and mounting systems plus bear the associated wind and snow loads

SOLAR THERMAL SPACE PLANNING, PLUMBING AND WIRING

Plan shortest pipe run (less than 15m or 50ft) from collectors to storage tank location, near conventional hot water tank

Install 2–19mm (3/4in) capped copper pipes (with ends above roof) and 2–24V wires (14-gauge shielded wires)

Provide extra space for a (future) storage tank, approximately 1200 by 1200 mm (4 by 4ft), immediately adjacent to the conventional water heater and 600 by 900mm (2 by 3ft) wall space for controls

Rough in a plumbing bypass for the solar system

Extra wall outlets and/or hardwired rough-in for heat transfer equipment in mechanical room

All wiring should have an extra coil at both ends of run and be labelled clearly

PV SYSTEM SPACE PLANNING, PLUMBING AND WIRING

Provide a 50mm (2in) conduit, or fish appropriately sized wires through to connect the solar array, the inverter, and the electrical panel, all wiring should have an extra coil at both ends of run and be labelled clearly

Provide a circuit in the breaker box for the solar electric feed

Provide a vertical wall area to mount a 1200 × 1200mm (4 × 4ft) inverter panel, in utility or mechanical room. Mount controls on 12mm (1/2in) plywood backing instead of using anchors in drywall

For a grid-connected system, install an electric service disconnect switch as required by local utility

For a system with batteries, a ventilated, self-contained closet, bench or other space that can be locked should be designed to be close to the mechanical room

If you need to use a racking system, you can pre-mount brackets prior to roofing (or re-roofing). Check the manufacturer's specifications for required bracing.

CONTROLLING CLIMATE

In a passive solar design, your first step in climate control is site design – modifying your site to improve its exposure to the sun, shield against winter winds, capture summer breezes and otherwise moderate the climate outside your house. Your second step is the house itself, as it supplies both heating and cooling to the space it encloses. Heating and cooling are based on simple thermal properties, such as solar gain, heat movement, heat loss and thermal storage.

For your solar house to perform properly, it must capture the sun's energy, and control the captured energy so the interior climate remains comfortable: even temperatures, no drafts, comfortable humidity levels and good ventilation. It is easy to 'design in' features that result in overheating, moisture problems or indoor air pollution. All three are intolerable extremes that can be difficult, expensive or even impossible to control. Your design needs to allow for vents, fans, ducts, thermal mass and other means of control to modify the interior climate and give you flexibility. In short, you design out problems and design in controls and flexibility.

Not every room or space requires the same climate. Living and storage spaces have different climate requirements, as do greenhouses, attached sunspaces or seasonal solariums. Once you know the requirements of the different areas in your house, you can design the controls to create suitable climate 'zones'.

solar gain

Solar gain is the amount of radiation from the sun that reaches your house and is captured by glazing, solar thermal collectors, or a PV array. The amount of solar radiation that reaches the Earth is affected by both the atmosphere and cloud cover. The incoming radiation is broken into direct, diffuse and reflected components. Direct radiation is the most intense component and is the largest amount of radiation we receive on a clear day. When direct radiation is scattered by dust, air molecules and clouds it becomes diffused radiation. Reflected radiation is most prominent on a clear day and requires a surface (like snow cover) to reflect (rather than absorb) some of the direct radiation incident on the surface.

The radiation we receive from the sun comes to us in the form of short wave radiation, caused by the extreme heat of the sun. When these waves reach an opaque surface – i.e., a surface that is not transparent – they are either absorbed or reflected. Whether an opaque surface absorbs or reflects short wave radiation is dependent on the colour and texture of the surface itself. Most of the radiation we receive from the sun is within or near to the visible spectrum of light – ultraviolet and infrared are at the furthest ends of the visible spectrum. What we perceive as colour is actually a portion of radiation that has been reflected off a surface. The radiation absorbed by a material is converted into thermal energy, which we call heat. This means an object or material with a white surface, which reflects most of the radiation, will remain cooler than an object with a black surface, which absorbs most of the radiation. How much radiation a surface absorbs or reflects is also dependent on the texture of the surface. A shiny polished surface, such as a gray car roof, will reflect more radiation than the roughened surface of a gray concrete slab.

A surface that reflects visible radiation in the yellow-band of the spectrum and absorbs all other radiation looks yellow. A surface appears white when it reflects most of the visible radiation and black when it reflects very little of it.

HOW RADIATION FROM THE SUN REACHES US

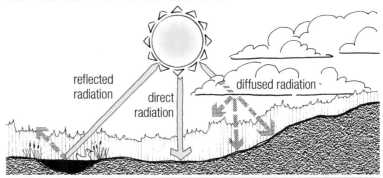

reflected radiation
direct radiation
diffused radiation

Your solar house takes in all three components of the sun's energy when sited properly. This means that even on an overcast day, your house will still have some solar gain.

HOW SURFACES ABSORB OR REFLECT RADIATION

R = Red Light Y = Yellow Light B = Blue Light

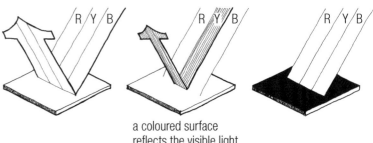

a white surface reflects most of the visible light spectrum

a coloured surface reflects the visible light in the spectrum of that colour and absorbs the rest

a black surface absorbs most of the visible light spectrum

Short wave radiation is absorbed or reflected by opaque materials, but it is transmitted by translucent or transparent materials such as glass. This means a large amount of solar energy passes through most materials used for glazing. The remainder is either reflected or absorbed by the glass. Objects within the glazed area absorb a portion of the short wave radiation and convert it to thermal energy which in turn radiates, in the form of long wave radiation, what we call heat.

Since long wave radiation does not pass through glazing as readily as short wave radiation, much of it stays inside the glazed area as heat. This is known as the greenhouse effect. A solar home relies on the greenhouse effect for heat collection, using south-facing glazing to admit sunlight, and dark-coloured walls or floors to absorb and transform the sunlight into heat energy. Glazing is responsible for the heat gain and light levels in your house. However, poorly designed, installed or situated glazing results in serious heat loss.

The key points in good passive solar design for cold climates are to maximize solar gain and minimize heat loss in the winter, and, to minimize the solar gain and maximize cooling in the summer.

THE GREENHOUSE EFFECT

Short wave radiation from the sun passes through glass to warm the surfaces of objects inside a space. The long wave radiation from the warmed surfaces does not pass through the glass easily, so most of the radiant heat stays in the space. Heat loss is due to conduction.

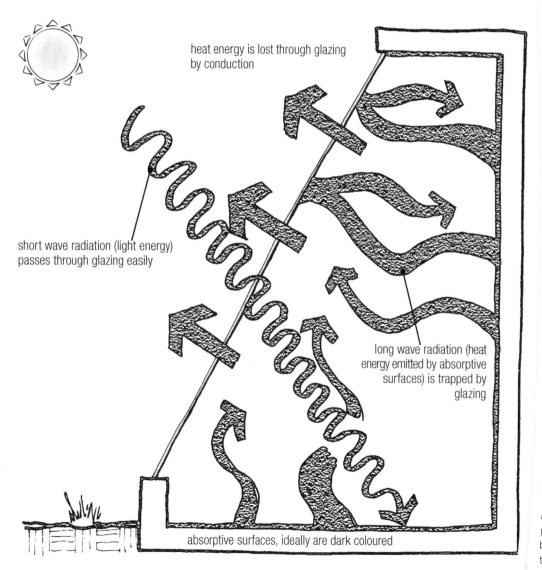

heat energy is lost through glazing by conduction

short wave radiation (light energy) passes through glazing easily

long wave radiation (heat energy emitted by absorptive surfaces) is trapped by glazing

absorptive surfaces, ideally are dark coloured

Although your glazing is responsible for the solar gain in your home, it is also responsible for a large portion of your house's heat loss. This is because glass is poor insulation material and conducts heat readily. However, more heat is lost through radiation than through conduction, especially at night when the sky acts as a near-perfect 'black body', absorbing heat from most objects on the Earth.

Heat loss can be reduced by installing low-e, double paned glazing in your house. Low-e glass enhances the greenhouse effect by blocking passage of the long wave radiation and conductive heat through the glass. Heat loss through conduction is also reduced by double paned glass. The airspace between the two panes acts as an insulator, blocking the flow of heat through the window.

properties of heat flow

Heat is a form of energy transferred between two objects when the objects differ in temperature. The greater the temperature difference, the greater the rate of heat transfer. Heat always flows from the warmer object to the cooler one until both are the same temperature, a state called thermal equilibrium.

All systems or objects that involve heat transfer strive to maintain thermal equilibrium, including our own bodies and the molecules that constitute the air around us. Thus air molecules are constantly transferring heat to each other. Temperature also affects the density of air molecules, so that warming molecules become less dense and lighter and tend to rise, while cooling molecules become more dense and heavier and tend to fall.

Heat flows in and out of a building at various rates, depending on both the insulating qualities of the materials used, and the way they are put together. Heat flow is a primary concern for a comfortable and healthy indoor climate. In a conventional home, most of the effort is focused on controlling the heat flows between inside and outside. An energy-efficient home is airtight and well insulated, minimizing the heat flow between inside and outside.

In a passive solar home, you have to minimize the heat flow between inside and outside but you also have to control heat flow within the building in order to avoid discomfort due to overheating, underheating or temperature stratification. Understanding how heat flows will help you plan your climate control system. Essentially, heat moves in four ways: radiation, conduction, convection or mass transfer.

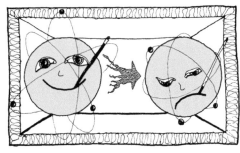

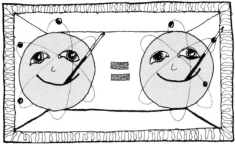

Molecules transfer heat until they reach a state of thermal equilibrium. A whole system, such as your solar house, works in the same way. For example, in the winter, heat is transferred from the warm interior to the cold exterior.

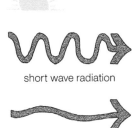
short wave radiation

long wave radiation

conduction

convection

temperature stratification

To maintain a comfortable climate in your solar house, heat losses and heat gains must balance.

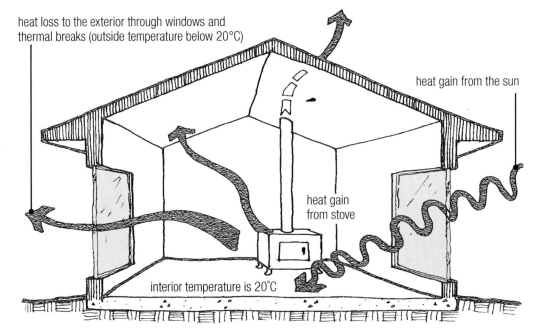

heat loss to the exterior through windows and thermal breaks (outside temperature below 20°C)

heat gain from the sun

heat gain from stove

interior temperature is 20°C

radiation

Short-wave radiation is the way the sun's energy comes to us on Earth. When it passes through glazing and is trapped in a house, the greenhouse effect results in long energy waves moving from a warm surface to a cooler surface. Like light from a light bulb, radiation is directional, moving through the air until it strikes a solid object. We can feel a wood stove's radiant heat from quite a distance as long as no solid objects block its path – the wood-stove effectively heats what it can 'see'. Radiant heat is often used in large buildings such as arenas because it heats solid objects, like hockey fans, before it heats the air, allowing us to be comfortable at lower air temperatures. One of the reasons in-floor radiant heating systems have become very popular in new housing is that the low-temperature heat radiated from the floor keeps your feet warm, which means that you feel warmer in general, even at a lower air temperature.

Our bodies are a source of radiant heat as well. We interact thermally with walls, windows and other objects that make up our house's environment. This explains why we feel cooler when close to a window or other cold surface. On a cold day or night, the temperature of a window surface is several degrees below both room temperature (20°C/70°F) and our own body temperature (37°C/98.6°F). Because heat flows from warm to cold surfaces, we radiate heat to the window surface and this makes us feel cold. The colder the window surface, the faster the rate of heat flow from us to the window. This works in the opposite way as well: we feel hot when close to a lit stove or fireplace, even though the overall room temperature remains the same. The hotter the source, the more intense the heat flow is to us.

If we can keep the inside surface of the window from becoming too cold, we will be more comfortable, because the rate of heat flow will be slowed down. We can use multiple layer glazing, which will insulate the inside window surface from the cold outside temperature.

RADIANT HEAT

We can feel the radiant heat from a wood stove from quite a distance as long as no solid objects block its path.
The low-grade heat from wood stoves and other heating appliances is actually long-wave radiation, like that of the low-grade heat that results from the captured solar gain in your house.

Like light from a light bulb, radiation is directional, moving through the air until it strikes a solid object.

All materials radiate energy at all times, but the amount of thermal energy a material radiates depends on the temperature of the radiating surface

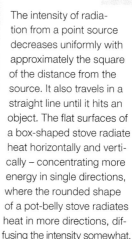

The intensity of radiation from a point source decreases uniformly with approximately the square of the distance from the source. It also travels in a straight line until it hits an object. The flat surfaces of a box-shaped stove radiate heat horizontally and vertically – concentrating more energy in single directions, where the rounded shape of a pot-belly stove radiates heat in more directions, diffusing the intensity somewhat.

OCCUPANTS AS A HEAT SOURCE

Along with appliances and other pieces of equipment that radiate heat to the inside air, we make up the 'internal gains' of a house. As we build better-insulated houses with tighter building envelopes, internal gains have a greater impact on the ambient temperature within the house, and play a larger role in comfort levels. In fact, there is a fabulous example of a row of terraced house development in Goteborg, Sweden, where the heating load for each unit is met solely by the occupants, the appliances and lighting, and the heat recovered by the ventilation system. These houses are well-insulated, airtight buildings with good passive solar access and were part of an international monitoring programme.

without impeding solar gain or the transparency of the glass. As low-e coatings reduce ultraviolet (UV) radiation through a window, they also reduce potential fading of upholstery and carpets.

Radiation works together with the next form of heat transfer, conduction, to accelerate the heat flow through walls, windows and roofs. When the surrounding terrain is colder than the outside surfaces of your house, there will be a net flow of thermal radiation from the house to its surroundings. Your roof also radiates substantial amounts of energy to the cold night sky. Nighttime cooling is good in the summer, but not in the winter, which is why a well-insulated roof is necessary to prevent heat loss.

Radiant heat can be absorbed or reflected by an object. Whether an object reflects or absorbs radiant heat energy depends on its surface and its temperature. To absorb radiant heat energy, an object must be relatively cool, be a good thermal conductor, and have a surface which is non-reflective and absorbing, such as flat black. Most common building materials, such as wood and masonry, have a relatively low value of reflectance and so are good heat absorbers. To reflect radiant energy, an object must be a poor conductor, with a surface that is smooth, polished and reflective, such as shiny metal foil. Aluminum foil is often used on one or both sides of some insulating materials, as it reduces the overall rate of heat loss (or gain).

Nearly half of the heat loss out a double-pane window is due to radiant heat loss. A low-emissivity (low-e) coating reflects radiant heat back into the house in the same way as aluminum foil

Window coverings can slow the rate at which heat is radiated to windows, however, they can also cause condensation problems. Forced-air heat registers must be placed in front of drapes or curtains. Otherwise, the heat flowing behind the coverings will heat the windows, which in turn will radiate the heat outdoors.

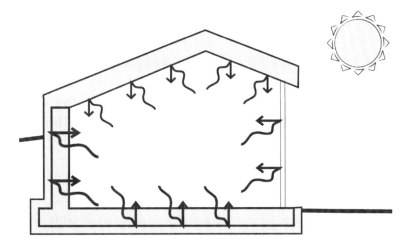

conduction

Conduction is the movement of heat through solids. It can be visualized by the way heat moves from the surface of a stove through the cooking pot and into its contents. The excited (hot) molecules of the surface of the stove element or burner transfer some of their energy, in the form of vibrations, to the molecules of the cooking pot, which in turn, transfer energy to the contents of the pot. Some of the energy is also transferred to the handle, which then conducts some of its heat energy to your hand. We often learn about conduction early on in life by grabbing hold of the hot handle of a cooking pot and burning our hands.

The rate of heat movement depends on the material's ability to conduct heat. If the material has little resistance to heat flow, we call it a good conductor. If the material doesn't conduct heat well, it is resistant to heat flow and is called an insulator. Insulators have high 'R' (for resistance) values. Metals are good conductors. Glass is also a good conductor, which is why the major part of the heat loss from your home 'goes out the window'.

Poor conductors make good insulators. Insulation works by trapping small pockets of air which act as barriers to the transfer of heat through solid material. The more air pockets, and the more convoluted the path through those air pockets, the better the insulation. Fibrous insulations trap air in pockets between their fibres. Foam insulation has little cells of gas trapped in it. Although glass itself is a poor insulator, when used in two or more layers that trap one or more air pockets, its insulating value improves significantly. Injecting an inert gas such as argon or krypton between layers of glass increases the insulating value of glass even more by slowing the air current (convection) in the space between the glass panes.

Another component of conduction is the area and thickness of insulation materials. The rate of heat flow is proportional to the area through which the heat energy moves. This means that the larger the surface area of your house, the greater your heat-loss potential (think of water flowing through a hole – the bigger the hole, the greater the volume of water flowing through it). By using materials with high insulating values, we can slow down the rate of conductive heat loss.

Insulation works by trapping air. The more air pockets, the better the insulation. The trapped air slows the rate of heat transfer.

fur as insulation: air trapped between hair

styrofoam insulation: air trapped in foam

fibreglass insulation: air trapped in fibres

increasing the number of air spaces between layers of

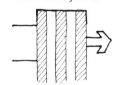

glass increases the insulating value of the window

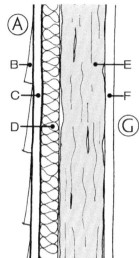

Each element that makes up a wall or roof structure has an insulating value, including a film of air on the interior and exterior surfaces. When added together, they indicate how resistant the wall or roof is to heat loss through conduction. Cross-section of an exterior wall, showing:

A = outside air film
B = siding
C = sheathing
D = insulating sheathing
E = batt insulation
F = drywall
G = inside air film

A + B + C + D + E + F + G = total insulating factor

See Appendix C for common building materials and their insulating values, Appendix D for heat loss calculations.

convection

Warming air rises as it expands, becoming lighter and less dense. Cooling air falls as it contracts, becoming heavier and more dense. This constant heat-triggered movement of gases and fluids is called convection. It is similar to conduction, but involves heat flows through fluids and gases in contact with solid objects or surfaces.

A convection loop occurs in a house when the air against a warm surface rises as it is heated, while the air against a cold surface falls as it is cooled. Convection loops occur at various scales and velocities in your house (see the illustration).

Convection results in spaces heating from the top down and cooling from the bottom up. This can affect your comfort, as moving air feels cooler than still air. Strong convection loops can keep you cool in the summer, but a draft of cold air across your feet in the winter is uncomfortable. Removing or isolating the cold surfaces in a room minimizes convection loops.

Tall window areas, such as you might find in a two-storey south-facing solar home, offer long stretches of cold surface, and a lot of cold air tumbles down the glass to your feet when the sun is not shining. This is why forced air heat

registers are placed under windows: the heated air counteracts this convective heat loss by intercepting the cold air downdraft from the window, preventing a whole-room convection loop. As windows have improved, this becomes less of an issue.

In well-insulated houses, where exterior wall and ceiling surfaces are kept relatively warm, convection loops are diminished. However, the lighter, warmer air is still at the ceiling, while the cooler, heavier air is at the floor. This layering of air temperatures is called 'stratification'. Areas with high ceilings tend to stratify dramatically. The temperature differences between floor and ceiling can be up to 15°C (27°F) in a room with a cathedral ceiling.

To avoid temperature stratification, most solar homes incorporate forced convection, supplied by fans and blowers with ducts. By using a fan or blower with a high-efficiency variable speed motor, you gain more control over the rate at which convection takes place.

In the winter, convection occurs between the inside and the outside of your house as well when the outer surfaces are warmer than the cold surrounding air. The heat from the wall is transferred at a faster rate when the wind is blowing because the wind creates a forced convection current. As with conduction, the greater the surface area in contact with the air, the greater the rate of convection heat transfer.

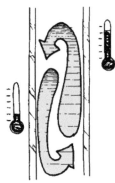

A convection loop can also occur between two panes of glass in a window unit. Convection can be slowed down by filling the airspace with a heavy gas such as argon.

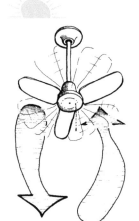

Forced convection – using fans and blower units to move air around – evens out temperature stratification

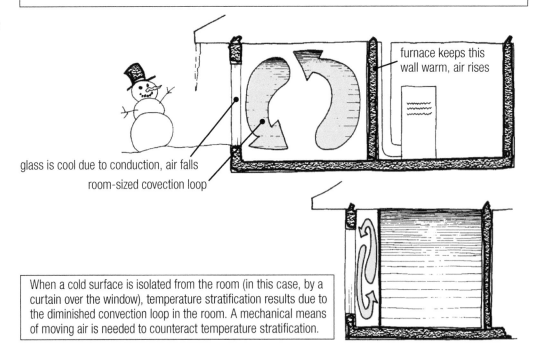

CONVECTION LOOPS OCCUR AT DIFFERENT SCALES WITHIN A HOUSE

furnace keeps this wall warm, air rises

glass is cool due to conduction, air falls

room-sized covection loop

When a cold surface is isolated from the room (in this case, by a curtain over the window), temperature stratification results due to the diminished convection loop in the room. A mechanical means of moving air is needed to counteract temperature stratification.

mass transfer

Mass transfer occurs when a material containing heat is moved. Remove the cooking pot from the stove, remove the energy. Mass transfer occurs when you remove heated air from the ceiling with a fan, or when you pump heated water from a solar collector to a storage tank. When mass transfer of air occurs from inside and outside through cracks and holes in walls, we call this exfiltration and infiltration. Infiltration and exfiltration can be controlled by making the building airtight and then exchanging air by mechanical means, such as vents and fans.

Although you can move a lot of heat with a little water, you have to move a large volume of air to move any significant amount of heat. This means, when you are designing air handling systems, you must size ducts and other elements to handle the amount of air being pushed through them.

HEAT LOSS

Reducing heat loss in winter, and heat gain in summer, is the best way to minimize your heating or cooling costs. You can reduce heat loss significantly by building to energy efficient standards such as the R-2000 program, or go even further and aim for Net Zero Energy or the European 'PassivHaus' standard (see the section on building envelopes for more about these standards). An energy-efficient home reduces the amount of heat required to maintain a constant temperature through high levels of insulation, airtight construction techniques, controlled ventilation and an efficient heating system. A solar house will not function effectively unless it is designed to contain the collected heat – otherwise, it defeats the purpose of capturing the sun's energy in order to conserve non-renewable fuels. In other words, an energy efficient house is not necessarily a solar house, but a solar house must be energy efficient.

Heat loss from leaky door frames, single-glazed windows and under-insulated areas of your home will cost you more, in the long run, than the combined initial costs of energy efficient construction, multiple-glazed windows with low-e coatings and high levels of insulation.

HEAT LOSS THROUGH WINDOWS

Heat is lost through glazing by conduction, radiation and convection, and by air infiltration around frames. Low-e coatings effectively slow radiant heat loss without blocking much visible light or solar heat gain. They also block damaging UV rays.

DOUBLE PANE UNIT

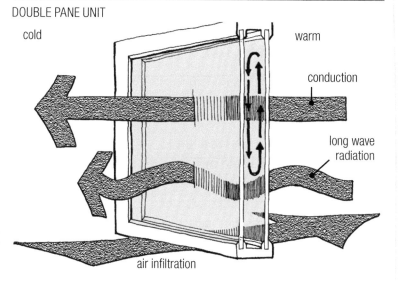

cold

warm

conduction

long wave radiation

air infiltration

DOUBLE PANED, LOW-E UNIT

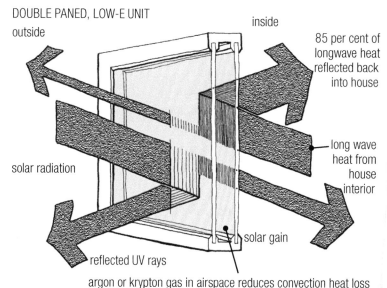

outside

inside

solar radiation

85 per cent of longwave heat reflected back into house

long wave heat from house interior

reflected UV rays

solar gain

argon or krypton gas in airspace reduces convection heat loss

Windows should face within 30° of true south if they are to act as effective solar collectors. Windows to the east allow for morning light while west facing glazing can lead to summer over-heating. Your design should include as few north facing windows as possible.

Properly sized overhangs can control summer overheating by shading windows that face to the southwest and southeast. Southwest and west facing windows can cause over-heating in the late afternoon and evening in the summer, when the sun is low in the sky but still powerful.

More information on climate zones can be found at www.sis.agr.gc.ca/cansis/nsdb/climate/hardiness/intro.html

glazing

Glazing is an integral part of climate control. Windows act as solar collectors, but they are also responsible for 15 to 35 per cent of the total heat loss in a house. The performance of your windows will largely determine how much of your solar gain is retained to become usable heat energy.

Four narrow bands of radiation affect windows: ultraviolet, visible and near-infrared (all relatively high energy, short-wave radiation), and far-infrared, the long-wave radiation given off by room-temperature objects, people, appliances and heating systems (the 'internal gains' of a building). Ordinary glass is partially transparent to the first three wavelengths, but not the fourth. About 66 per cent of the heat lost through glazing is lost by radiation. Conduction accounts for most of the heat loss across the spacer material separating the two glazing layers at their edges, and through the frame of the window. Air movement in the spaces between glazing layers causes convective heat loss. In addition, air leakage between moveable or operable frame components results in heat loss.

The style of window will also have a bearing on its ability to capture solar heat. A window with a wide frame and internal grille, or a window made up of several smaller units, has less glazing area available to capture solar energy than the same size thin framed unit with a clear glazing area. The second window style has a greater proportion of glass to frame area, allowing more sunlight into the living space.

If you are building a house that has little thermal mass, then you should limit your south facing glass area to no more than 10 per cent of the floor area that it sees. This will reduce the chance of overheating. If you are building a house with high levels of thermal mass (such as a slab on grade that is exposed to direct solar gain), the glazing can be increased to a maximum of 20 per cent of your floor area. Heat storage, glazing area and heat distribution must be carefully designed and balanced to avoid overheating.

In 'superinsulated' houses, it has been found that the amount of solar gain needed to meet the small heating loads required by the house goes down, so the maximum amount of glazing can be reduced as well.

A high-performance window may cost more than a conventional double-paned unit, but the payback more than justifies the expense. High-performance windows are a cost-effective route to energy efficiency. If oriented within 20° of south, high-performance windows have an average net energy gain over a 24-hour period in midwinter. In addition, the overall thermal comfort level in your house will improve, you will 'gain' floorspace because the area directly in front of the windows will be more usable, and the noise transmission from outside to inside will be diminished.

Window Energy Ratings are a much more accurate indicator of energy performance than the U-value alone, because they take a range of factors into account, including the useful solar heat gain. Under most systems, a positive value indicates a high-performance window, while a negative value indicates a poor choice in terms of energy conservation.

The British Fenestration Rating Council (BFRC) Scheme is recognised as a way to comply with UK Building Regulations (Part L) for new or replacement windows. This system is the basis for a major EC initiative to create a window energy rating system.

In the US, the National Fenestration Rating Council (NFRC) operates a voluntary programme for energy performance ratings. The NFRC label can be found on all ENERGY STAR® qualified window, door, and skylight products, but ENERGY STAR® currently bases its qualification only on U-factor and SHGC ratings.

Search for current websites using keywords: BFRC, NFRC, energy star, windows, window rating systems.

Consider exterior shading devices, as its too late to deal with overheating once the sun hits your windows.

See the section on building envelopes (page 122) for details on window types and glazing options.

thermal mass

Passive solar design is an approach that integrates building components – exterior walls, windows and building materials – to provide solar collection, heat storage and heat distribution. New construction offers the greatest opportunity for incorporating passive solar design features into the structure of the building. For retrofit projects, the best ways of incorporating solar are daylighting strategies, heat control techniques, and using passive solar heating and cooling strategies to allow modification of mechanical systems.

Passive solar designs are typically categorized as sun-tempered, direct-gain, indirect gain and isolated gain. Sun tempering is simply using windows with a size and orientation to admit a moderate amount of solar heat in winter without special measures for heat storage. Direct gain has more south-facing glass in occupied spaces and thermal mass to smooth out temperature fluctuations. Indirect gain designs put the thermal mass directly behind the glazing to reduce glare and overheating in the occupied space. Isolated mass designs, where the thermal mass is isolated from the collectors, add significantly to construction costs without a counterbalancing contribution to the overall efficiency of a house. At one point isolated systems where air flowed over rock storage bins

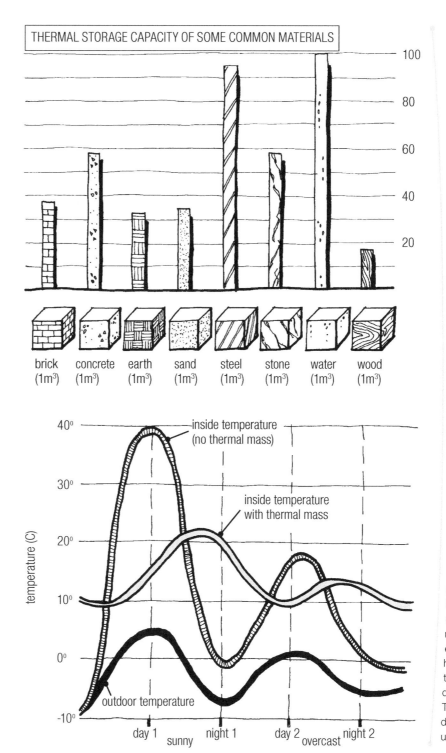

THERMAL STORAGE CAPACITY OF SOME COMMON MATERIALS

brick (1m³) concrete (1m³) earth (1m³) sand (1m³) steel (1m³) stone (1m³) water (1m³) wood (1m³)

inside temperature (no thermal mass)

inside temperature with thermal mass

outdoor temperature

temperature (C)

day 1 sunny   night 1   day 2 overcast   night 2

THERMAL MASS AND TEMPERATURE MODERATION

Thermal mass moderates temperature extremes in your solar house by absorbing excess heat during sunny days and releasing it overnight and during overcast days. The chart at right shows both the moderating effect thermal mass has on temperature, and the 'flywheel effect', which delays temperature extremes. The length of this time lag is dependent upon the material used for thermal storage.

were seen as one way of working with passive solar. Better window technology and more cost-effective methods of creating energy-efficient homes have made this option less favourable. Although not a passive solar system the current 'best' isolated gain system to include in any home is a solar hot water system.

Thermal mass is used to store surplus daytime heat for night or cloudy day use. It will also moderate and delay temperature extremes daily and seasonally in your home. Thermal mass is any material used to store heat: water, masonry materials such as concrete, brick and ceramic tiles as well as gypsum board and other building materials.

Thermal mass needs to be designed to do two things: first, it needs to quickly absorb solar heat for use over the diurnal (day-night) cycle and to avoid overheating, then it needs to slowly release the stored heat when the sun is no longer shining. Depending upon the local climate and the use of the building, the delayed release of heat may occur a few hours later or slowly over days.

Venting can be another solution for handling stored heat, eliminating late afternoon heat when the building's thermal mass is already saturated. Venting can also be used as cooling, bringing outside air into the building when it is cooler than the building's thermostat setting. Venting requires an exhaust fan tied to a thermostatic control or flushing through natural ventilation (i.e., open the window!).

The thermal capacity of a building can be improved by increasing the thickness of the gypsum board used on interior wall surfaces of the building or by using thicker gypsum board products. Increasing the thickness of all of the wall surfaces in your home can raise its thermal capacity for little additional material cost and practically no labour cost. It has the added benefits of increasing the fire safety and acoustic privacy of interior spaces. This diffuse thermal mass approach depends on effective convective airflows since room air is the heat-transfer medium. To really 'charge' the walls, temperatures within the space must be allowed to fluctuate about 3°C (5°F) above and below the thermostat setting. With a typical thermostat setting of 21°C (70°F), this gives a temperature swing of 18–24°C (65–75°F) – hardly a challenge! However, most electronic thermostats now have a smaller allowable swing than this.

material/thickness	direct gain m² (ft²)	indirect gain m² (ft²)	isolated gain m² (ft²)
concrete			
10cm (4")	0.4 (4)	0.7 (7)	1.3 (14)
15cm (6")	0.3 (3)	0.5 (5)	1.3 (14)
20cm (8")	0.3 (3)	0.5 (5)	1.4 (15)
brick			
5cm (2")	0.7 (8)	1.4 (15)	1.9 (20)
10cm (4")	0.5 (5)	0.8 (9)	1.7 (18)
20cm (8")	0.5 (5)	0.9 (10)	1.8 (19)
gypsum board			
1.8cm (0.5")	7.1 (76)	10.6 (114)	10.6 (114)
2.5cm (1")	3.5 (38)	5.3 (57)	5.3 (57)
hardwood			
2.5cm (1")	1.6 (17)	2.6 (28)	3 (32)
softwood			
2.5cm (1")	2 (21)	3.3 (36)	3.6 (39)

Cement board or old plaster has better heat transfer properties than gypsum board because of the paper backing

The chart at right shows the square metres of mass needed for each square metre of south facing glazing. Imperial measurements are in brackets.

While water provides good heat storage – twice that of most masonry material – concrete is often used because it is usually already part of the structure and does not add to the cost.

direct gain

Direct gain can be the easiest, and most cost-effective way to incorporate solar into your design. The placement of your thermal mass is important. It should receive as much direct sunlight as possible during the winter months, and be shaded in the summer, to avoid overheating. Thermal mass must be exposed to the interior of the space, and insulated from the exterior. Most houses will have sufficient thermal mass in their foundations to store the collected heat. To be effective, though, this mass must be a non-reflective, dark coloured surface in the living space, not in the basement.

Thick, well-insulated walls and floors of concrete, stone or brick

> ### DIRECT GAIN CONCRETE SLABS
>
> A concrete slab will absorb heat to a depth of approximately 10cm (4in). However, the most usable heat lies in the first 1.25cm (1/2in). By placing ductwork in the slab, you can activate the remaining masonry more readily. Solar gain is converted to heat and absorbed by the top part of the slab. When the temperature in the space cools, the slab releases the heat. Without a method of moving the heat around, temperature stratification can become a problem. An activated slab (right) can distribute heat from the sun (or any source) throughout the house.

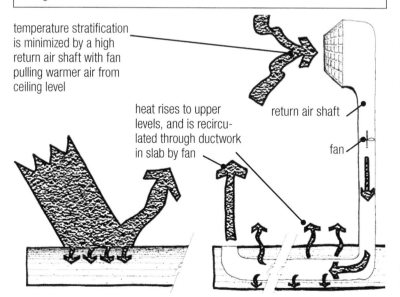

temperature stratification is minimized by a high return air shaft with fan pulling warmer air from ceiling level

heat rises to upper levels, and is recirculated through ductwork in slab by fan

return air shaft

fan

can be covered with dark tile or slate (any earth colour, but flat black works best) to improve collection efficiency. If you use a light coloured tile or slate, you will still collect heat, but not as efficiently as with a dark colour. Wall-to-wall carpeting and wall panelling block solar energy from reaching the mass and act as insulation between the mass and the living space. Walls tend to be more effective as thermal mass, because the vertical face of a wall can easily be tied into a convection loop, making the stored heat more accessible. The way you want to use the space within your home may limit the amount of exposed thermal mass for direct gain.

Direct gain thermal mass not only reduces the temperature range in your home; it also delays the time at which temperature extremes occur. Thus the sun's energy is stored for later use when the sunlight is gone. Thermal mass also moderates temperature fluctuations, as the heat it absorbs during the day is slowly released through the night, when outside temperatures are cooler.

A totally 'passive' solar home with a direct gain system relies on natural convection and conduction to move the heat around. Other than the installation of manually operated vents and windows to provide air circulation, there are no climate controls. This system works well for greenhouses and small spaces, but once you begin breaking the space up into rooms, you need a way to control the temperature and distribute the stored heat to other areas that don't have south-facing glass.

One way to distribute the heat is to 'activate' the thermal mass in some way. Slab-on-grade foundations can be activated by running a network of heating ducts through the concrete. A fan blows sun-warmed air through the ducts. The heat is conducted to the slab from the ducts, and gradually radiates back into the living space. This type of system also works well with a wood-stove as an auxiliary heat source, as the heated air is distributed and radiated in the same way.

CONCRETE MASS

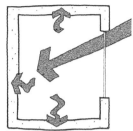

daytime: gain through radiation and conduction

nightime: loss through radiation and convection

WATER MASS

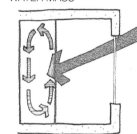

daytime: gain through radiation and convection

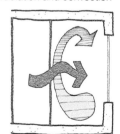

nightime: loss through conduction and convection

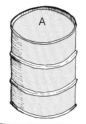This arrangement of small
water containers holds the
same amount of water as
the large drum, but has five
times the surface area. This
means the small containers
heat up and give back their
stored heat more quickly.
They are useful for controlling quick temperature
changes. The large drum
will take longer to heat up
and longer to give back its
heat. This long term storage
regulates and moderates
temperatures overnight.

volume A = volume B

5 × surface A = 1 × surface B

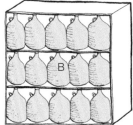

*Concrete/rock slabs should be 100–200cm (4–8 inches) thick, and
you should allow 1–2 cubic metres per square metre (c.f. and s.f.)
of glazing. A rule of thumb: at least one half to two thirds of the total
surface area in a space should be constructed of thick masonry.
For water storage, total mass should be 100 to 200 litres per square
metre of glazing (2–4 Imperial gallons per square foot).
Long and short term water storage is most commonly used in
greenhouses and solariums, as most people don't want water barrels
cluttering up their living space. Long term and short term storage
is dependent on the surface to volume ratio of the containers. Heat
is absorbed and released more easily if the thermal mass has a large
surface area compared to its volume. See illustration at left.*

Hydronic in-floor heating systems with solar thermal providing a good portion of the space heating requirement can also
be incorporated into a concrete slab. This type of system is
perfectly suited to maximizing the solar potential in houses
with less-than perfect sites for passive solar gain – for example,
tight urban lots, north-facing slopes with great views, tall
houses in forest clearings.

The biggest challenge with in-floor heat is combining it with passive
solar. Careful design is necessary to ensure that the systems are
not competing with each other or cancelling each other out. *See
the section on auxiliary heating systems for more details.*

Interior water walls can be used as thermal storage in a similar manner to concrete slabs or walls. The storage is usually
contained in one south-facing wall, so that the sun's rays hit
it for most of the day. During the day, the mass is heated. At
night, as the outdoor and indoor spaces begin to cool, the heat
is radiated back into the space. These walls are usually made
of plastic or metal containers, such as acrylic 'suntubes' or
painted 45-gallon drums. Water storage containers must be
carefully maintained – any leaks can cause major damage to
your house and its contents. They should be placed in drainage
pans fitted with a floor drain.

indirect gain

An indirect system is where the thermal mass is located between
the sun and the living space. For example, a sunspace keeps the
glass and mass separate from the occupied space but allows for
the transfer of useful heat into the building by convection or a
common mass wall; temperatures in a sunspace are allowed to
fluctuate around the comfort range.

These systems are more complex than direct gain systems, as the
heat absorbed by the mass must then be transferred to the living
space. Thermal storage walls can use masonry or water as mass.
Masonry, stone or concrete walls transfer heat primarily by conduction. Water thermal storage walls transfer heat through the
wall mainly by convection.

Glassed-in porches, solariums and greenhouses can be used as
indirect systems, with high and low vents transferring heat
between the sunspace and the living space. With their high ratio
of glass to floor area, they may require additional heat storage.
This can be accomplished by adding mass to the foundation or
by installing additional thermal mass in the interior (masonry
knee walls or water storage).

Photo courtesy of SHIP Database.

The amount of radiation available to indirect gain spaces can be increased by placing a bed of white stone or shell in front of the exterior wall, or by situating the wall in front of a swimming pool or other body of open water. Radiation hitting the reflective surface is bounced back onto the storage wall, which is especially effective on cloudy days when the vertical surface will not be receiving much direct radiation. Proper shading must be incorporated to prevent the collectors or glazing from reaching high temperatures in the summer.

The major disadvantage of indirect gain designs is that the amount of window space available on the south face of the house is limited. But this is not completely negative – mixing direct and indirect gain systems can boost the solar component of your house. Where solar gain is good but views are not great, or noise from a busy street is a concern, indirect gain systems can be the best solution. For more information on the sizing and construction of indirect gain designs, please see the resource section, (page 196).

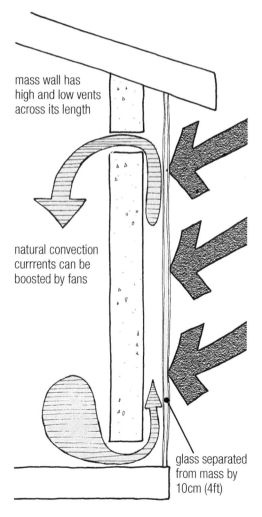

mass wall has high and low vents across its length

natural convection currrents can be boosted by fans

glass separated from mass by 10cm (4ft)

THE TROMBE WALL
Named after Felix Trombe, the French scientist who developed it, the Trombe wall is typically a south-facing glass wall with a massive masonry wall (painted flat black or another dark colour) located 100mm (4in) or more directly behind the glass. Solar radiation passes through the glass and is absorbed by the outer face of the masonry wall. The heat is then transferred through the wall by conduction, and from the inside face of the wall to the space by radiation and convection. High and low vents in the mass wall allow the flow of natural convection currents. These currents move the heated air from the airspace between the wall and the glass into the room.

isolated gain

Isolated gain, or remote storage designs, have the solar collection and thermal storage isolated from the living space. The system functions independently of the living space, only drawing heat from the system when it is needed. Probably the best use of an isolated gain system is as a solar domestic hot water system. Space heating with isolated gain systems can be complex and expensive. With the storage isolated from where the heat is required, pumps and other mechanical devices are necessary to pump the heat, leading to lower efficiencies. However, in-floor radiant space heatng systems can benefit greatly from a carefully designed solar hot water system, especially if direct solar gain is not an option for your home.

Photo courtesy of Conserve Nova Scotia

Solarwall, a Canadian-made product, uses the Trombe wall concept to preheat ventilation air behind metal siding.

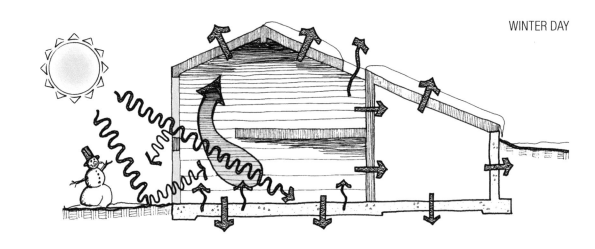

SEASONAL HEAT GAIN AND LOSS IN YOUR SOLAR HOME

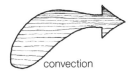

short wave radiation

long wave radiation

conduction

convection

temperature stratification

WINTER HEATING MODE

With its large area of south facing glass, most of the low sun's radiation is captured, passing deeply into the house with a lot of direct gain. The air recirculation is charging the slab, filtering the air, pulling in fresh air and driving the ventilation system.

At night, during the winter, the recirculation continues to destratify the temperature while reclaiming stored heat from the slab.

SWITCHING MODES

Because of high thermal mass, the house is very slow to react to changing outdoor temperatures. It takes a lot of heat to recharge a cold slab and it takes some time for a heated slab to cool. You will need to experience living with your home's thermal personality before you know when to switch from winter heating mode to summer cooling mode.

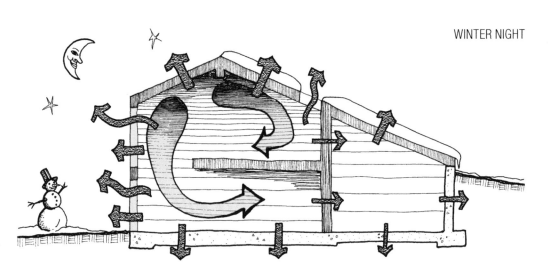

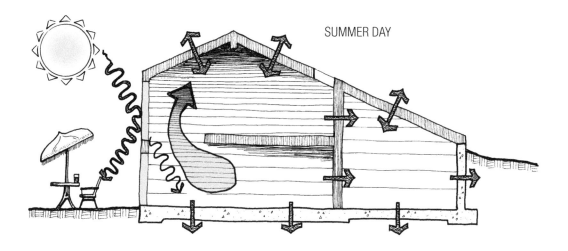

SUMMER DAY

SUMMER

With its large area of south facing glass, most of the high summer sun's radiation is reflected. What little passes through, does not penetrate deeply into the space and little of it is absorbed by the thermal mass through direct gain.

The air recirculation is off and the house relies on natural ventilation (opened windows). Also, no heat is being put into the slab. As the slab goes to near-ground temperature, it becomes a large cooling mass.

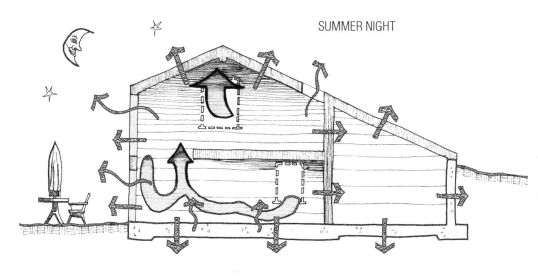

SUMMER NIGHT

COOLING MODE

To take advantage of the slab as a cooling mass, it is best to use minimal ventilation during hot days by cracking several lower windows and opening the highest windows or skylights. At night, maximize ventilation by opening vents and windows to ventilate the heat absorbed during the day to the outside. This cools the thermal mass, allowing it to absorb the heat of the next day.

Cross-ventilation is a traditional method of natural ventilation that works with the prevailing summer wind and the 'stack effect'. The stack effect works in conjunction with the principle of convection: warm air rising to the top of the house is vented and the resulting negative pressure pulls cool air into the house through low openings.

This can be accomplished by opening a high window or a vent on one side of the house and slightly opening low windows or vents on the other side of the house. The resulting increased air movement increases evaporation from the skin and makes you feel cooler and more active. Hot, dry air can be moistened by having the lower open windows or vents situated on a highly vegetated and shaded face of the house, or near a pool of water.

AUXILIARY HEATING

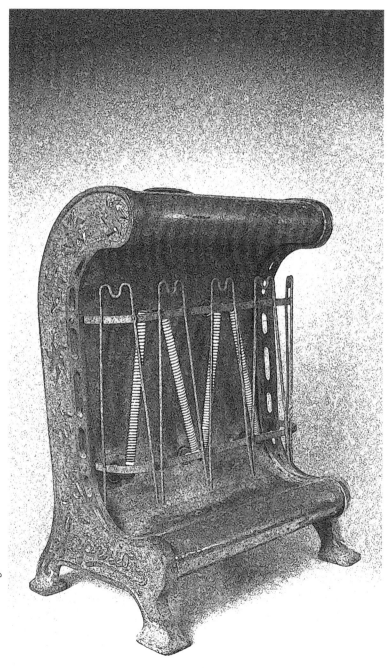

In cool climates you can use the sun to generate between 40 and 60 per cent of your space and water heating requirements. With little or no additional construction costs, you can increase this percentage. Depending on your budget and how you configure your home and heating systems. There are many variables determining your solar percentage: site orientation, solar access, the emphasis you put on solar capability in your design, the energy efficiency of your structure, your 'comfort zone' and your lifestyle.

While a totally solar house is technically possible, it may be more challenging to build. Depending on your builder, this may translate directly into 'more expensive'. However, we can move very close to a totally solar home in many cool climates with a super-insulated structure that includes the best possible windows, a well-planned solar thermal collection system, (possibly with seasonal storage capacity), and a change in your lifestyle. Regardless, you will need to install some type of back-up heat source for use during extended cloudy periods and extreme cold spells. This back-up heating will require a system to generate and distribute heat, and of course, a fuel source. It doesn't have to be a big system, or even one that would traditionally be considered capable of heating an entire house.

Maximize the building envelope and minimize the mechanical systems. Good quality controls and well-planned, simple systems can make your life easier. You might not want to be the first on your block to try newly introduced equipment...go for tried and true technology.

A back-up heat source can be any conventional residential heating system – forced air, hot water, or radiant heat – fired by electricity, oil, gas, propane or wood. The heat can be delivered to your living space via a furnace, a boiler or water heater, or a heat pump, installed in the same manner as in a conventional home. However, you will likely need a smaller than normal system. Water heating can be solar thermal with a back-up heater or boiler, space and water heating can be integrated into one unit. In some systems, the choice of system is up to you – however, it must be appropriately sized for an energy-efficient, airtight home, and it should complement your solar heating method (passive gain and/or solar thermal).

With lower heat losses and higher solar gains your home will need little additional heat. Heating equipment manufacturers are geared to the average housing market so their systems are sized to meet higher energy requirements. The output capacity of oil or gas furnaces starts at about 50,000 British Thermal Units (Btus), much higher than a super-insulated house requires, let alone a super-insulated solar home. With the exception of modular systems, this leaves you with fewer choices for a small-capacity heating unit, especially if you are considering a forced-air system.

The smaller your space heating load, the more important your other energy requirements become. Hot water appliances and lighting all require fuel sources and so, they all have potential for conservation. They can also provide you with a portion of – or all of – your back up space heating needs. Each occupant radiates about 75 watts of heating energy, and another 200 to 300 watts are supplied by appliances such as freezers, ranges and fridges. Lighting also provides a small amount of heat. In total, the average home generates 500 watts or more of the total energy required for space heating.

If you can find a small auxiliary heating unit, the low energy demand of your solar home can result in long-term capital cost recovery. In fact, the payback period for larger systems can be

A super-insulated solar house can have a heating load that is 70 to 90 per cent lower than a typically constructed house

- *Choose a system that is simple, affordable and easy to maintain. As the amount of purchased energy required to heat your home goes down, so does the importance (and the size) of the backup system. First and foremost, invest as much as you can in the building envelope. The heating system will be smaller, therefore less expensive, leaving more money for super-insulation and high tech window costs.*

- *Choose an integrated system instead of separate systems for space heat and hot water, air treatment (filtration and humidity control) and ventilation (fresh air). There are some gas-fired units available that do all of these functions, although they are currently oversized for a well-insulated house of modest proportions.*

- *Choose a system that suits your lifestyle and your locale. You may enjoy being directly involved in the process of keeping yourself warm. You may consider the time – and design considerations – for preparing, handling, drying and storing firewood worthwhile. On the other hand, you may want to set the thermostat and forget about it.*

longer than the life expectancy of the equipment, even when higher future energy prices and environmental costs are considered. So, how do you choose an appropriate back-up heating system? Some guidelines are listed above.

One thing to add into the discussion of heat delivery systems is the concept of 'exergy', which is the amount of energy required to supply heat at a specified temperature. Most conventional systems used to heat space and water burn fossil fuel at very high temperatures, yet we are uncomfortable at air temperatures over 27°C (80°F), and are quickly scalded by water that is over 55°C (130°F). With the delivery temperatures so low, look for systems or components that generate, or scavenge, low-temperature heat.

heat scavengers

You can recover heat from exhaust air and from wastewater with off-the-shelf units that are readily available.

Exhaust air heat is recovered by a heat recovery ventilator (HRV) or energy recovery ventilator (ERV) unit as described in the ventilation section. High-efficiency units can scavenge up to 85 per cent of the heat leaving the building as exhaust air. As your heating needs dwindle, this contribution becomes more important.

Drainwater heat recovery (DWHR) units are sections of copper pipe that tie into the main drain pipe of your house. The copper pipe is wrapped with small diameter copper tubing. When you have a simultaneous flow of water in and water out, such as a shower, the tubing scavenges between 40 to 60 per cent of the heat from the water draining from the shower. This recovered heat is transferred to the potable water in the tube then fed back into the shower line via the hot water tank. This is a nifty, simple and long-lived bit of technology that is proven to pay for itself within a couple of years depending on your household bathing preferences. A household with two people who take baths will have a much longer payback time than a family with two teenagers who take lots of showers.

Other heat scavenging strategies include locating your hot water tank on the main living floor, so that standing losses (the heat lost to the surrounding air while the tank stands idle but hot) contribute to the 'internal gains'. An electric tank can be plonked into a storage closet fitted with a drain pan and floor drain, while gas and oil units need to be placed where they have adequate combustion air and/or meet all other installation standards.

When you move to energy efficient lighting and Energy Star™ appliances, you lose some of the internal gains that would be associated with standard efficiency units. However, like the heat recovered from exhaust air, as your overall heating requirements go down, the internal gains play a more important role in heating your house. It's not that there are any more kilowatts of internal gains; it's that the proportion of the overall heating load they represent increases.

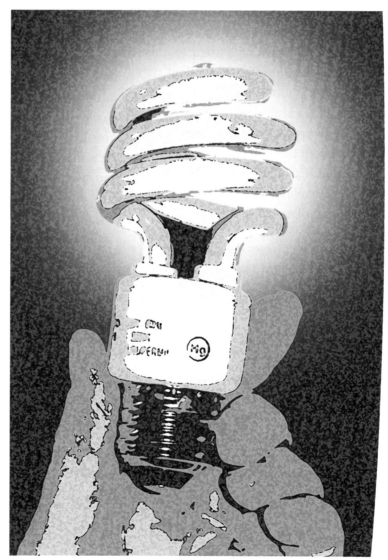

For more about drainwater heat recovery units, turn to the section on 'overall energy reductions' (page 125).
For more about ventilation and heat recovery, turn to the sections on 'controlling climate', (page 71) and 'indoor air quality' (page 99).

heat delivery systems

Forced air systems are inexpensive to install, as well as being reasonably easy to maintain and replace. They are able to collect and distribute heat from multiple sources located anywhere in your house, not just the burner. From this point of view, they work well with passive solar and/or woodstoves, or heat-exchange coils from a solar thermal system. During the shoulder seasons (fall and spring) where it is not too cold and you have decent solar gain, just running the furnace blower on low speed can destratify the air in your house and keep it at a comfortable temperature. They can easily be integrated with air filtration, recirculation and ventilation systems.

Wall and floor registers allow for flexibility in supplying heat to various locations, but also restrict furniture and drapery placement. Without proper filtration that is well maintained, dust, allergens and particulate matter can be blown through the house constantly.

Air systems must be designed into a home at an early stage since the ductwork can include large elements that are structurally and spatially integrated. In a well-insulated solar home, you may have a challenge finding a unit that is not vastly oversized for your heating load. Look for units that have high-efficiency burners, and also ensure that the model you are choosing has a high-efficiency, variable speed motor, preferably 'electronically commutated motors' (ECMs). ECMs increase the initial cost of the furnace, but reduce operating expenses.

Baseboard radiators can be either hot water or electric. They can provide a high degree of thermal comfort and can be easily zoned for different areas of your home. However, they can create problems with furniture placement and draperies. As well, they are awkward to place under patio or french doors. Hot water baseboards run at a higher delivery temperature and can also be under higher pressure than in-floor hydronics systems, so are not a good match to solar thermal, although evacuated tube collectors can provide adequate amounts of higher-temperature heat . A solar thermal system can be used as a pre-heater, especially if both

Forced air systems also include 'solar air' systems. Many of these units work on a room-by-room basis, and some models can be grouped together to provide a larger proportion of space heating needs. These units would be one way to get a higher solar contribution into a home with good solar orientation but where privacy is more important than a bank of windows. These systems require only a small duct through the exterior wall, and so are well-suited to solar retrofits. In addition to providing heat, they also provide ventilation air, although not all systems have substantial air filtration.

heating sources	advantages	disadvantages
electric	no combustion heater location flexible low maintenance cost low initial cost no fuel storage	expensive to operate thermostat limitations fuel supply cannot be changed easily baseboards require separate circulation & filtration placement of baseboards can be a problem baseboards 'fry' dust unless low-voltage thermostat used can be fossil fuel generated, increasing carbon footprint of house
oil	very efficient models available which need no chimney, can be combined with hot water supply	possible leaks in house – health hazard? maintenance needed combustion air required possible environmental degradation & pollution storage tanks
natural gas/propane	very efficient models available which need no chimney, can be combined with hot water supply	possible leaks in house – health hazard combustion air can be required storage tanks for propane
wood	works if power is off low fuel costs radiant heat contraflow heater/pellet stove innovations	manual labour storage space temperature/distribution control safety/combustion air required

space heat and water heat are supplied by the same boiler. Since they only provide heat, an air filtration, recirculation and ventilation system must be added. If you can minimize your heating load to practically zero, a few baseboard heaters might just do the job of taking you through the coldest parts of the year, especially in locations with extremely mild winters.

In-floor hydronic systems, also known as radiant floor systems, use low-temperature heat to create a radiant heat source at your feet, keeping you warm the same way that the sun heats objects on the Earth. Specially designed tubing can be embedded in concrete slabs, stapled under, or installed over, wood subfloors. To ensure optimal performance, installations under or over subfloors should include a metallic 'fin' of some type. These fins help spread the heat evenly across the floor, alleviating hot spots where the tubes run. Keeping your feet warm keeps the rest of you warm at a lower air temperature, reducing the amount of energy required to maintain a comfortable home. In-floor systems can be heated by any fuel source, including solar thermal systems. A well-insulated, solar home, with heat demand only during the coldest periods of the winter, could easily use a small water heating tank as the heating system, although boilers are often used. Since hydronics are under the finished floor, they do not impact on furniture or drapery placement. Because they only provide heat, an air filtration, recirculation and ventilation system must be added.

Solar thermal can be a fantastic match for in-floor heat, especially when passive solar gain is minimal due to site restrictions or orientation. You can orient your rooflines to take advantage of south and still have a solar home. On the other hand, if you have great passive solar gain, you may use your concrete slab as a thermal mass for heated air. Heat control can become a bit complex when using the same thermal mass for both your passivesolar system and your hydronic, in-floor auxiliary heat system. Both heating options rely on your floor slab to work,

but on different mechanisms. The slab is thermal mass for your passive solar gain, absorbing then releasing the heat from the sun. The slab is also a delivery system for your hydronic infloor heating, releasing heat into the space. If the hydronic system ignores the solar heat stored in the thermal mass, then you will have overheating and unnecessary fuel bills.

Ways of augmenting your passive solar with hydronic heat include scheduling your thermostats so that the hydronic system heats the house once the solar gain has dissipated from the slab. On a good solar day, this might be anytime after 9pm until 8am. This allows the slab to cool down prior to the optimum solar gain times between 10 and 2pm. Locating thermostats in the areas with best solar gain (but not directly in the sun's path), will key the thermostats to the rising temperature in the solar gain areas, so that the hydronic system stops supplying heat as the sun warms the space. Another option would be to run a loop from the direct solar gain area to a cooler portion of your house, one that doesn't receive any solar gain. This loop essentially 'robs' heat from the slab, reducing overheating. The design of these systems (and other options) is dependent on your particular home, and may require some energy analysis and modeling, and the assistance of a professional.

Heat pumps do not create heat through combustion – they simply move heat from one place to another. There are three types of heat pumps used for residential applications. Air-to-air heat pumps, air-to-water heat pumps and ground-source heat pumps (also known as geothermal or Earth-energy systems).

Air-source heat pumps draw heat from the outside air during the heating season and reject heat outside during the summer cooling season. They provide space heating and cooling, cooling-only or heating-only functions, but in most cases will also require an auxiliary heating source. A 'mini-split', or ductless, system are an excellent choice for retrofit options

In the main living space of a well-insulated passive solar home in most cool climates, a high-efficiency woodstove or a direct-vent gas insert can handle the bulk of the non-solar portion of space heating. Design your living space so that any chimney is placed on an interior wall, radiating heat into the living space. Mass walls located near any space heater will absorb and re-radiate heat as well. Your choice of space heater depends on your budget, the space to be heated and your choice of fuel.

in houses without existing ductwork. Up to eight separate indoor wall-mounted units, served by one outdoor compressor unit, can be installed in individual rooms of a house. They provide space heating and cooling, cooling-only or heating-only functions.

Air-to-water heat pumps are typically used in homes with hydronic heat distribution systems. During the heating season, the heat pump takes heat from the outside air and transfers it to the water in the hydronic distribution system. If cooling is provided during the summer, the process is reversed: the heat pump extracts heat from the water in the home's distribution system and 'pumps' it outside to cool the house. Air-to-water systems can provide domestic hot water as well as space heating.

A closed loop ground source heat pump (GSHP), uses the thermal energy of Earth as the heat source and heat sink for space heating and/or cooling. The heat is collected from the ground through a loop of underground piping in trenches or in U-shaped circuits in wells. An antifreeze or refrigerant solution, circulates through the piping, absorbing heat from the surrounding soil. An open loop geothermal heat pump uses the thermal energy of a body of water as the heat source and heat sink for hydronic system heating and/or cooling. Geothermal systems can provide water heating in addition to space heating and cooling. As the ground temperature is usually much more consistent than outside air temperatures in cool climates, GSHPs can be a more efficient choice than air-to-air units. They can also be used for seasonal heat storage, although you will need an energy engineer to help you optimize the system.

FUEL EFFICIENCIES
This table shows comparisons based on the amount of fuel required to provide one million British Thermal units (Btu) of useable heat energy (a Btu is the amount of energy required to raise one pound of water one degree Fahrenheit). One million Btus = 293 kilowatt-hours (kWh-e) = 81.5 megajoules (MJ).
Information from CANREN website www.canren.gc.ca/prod_serv/

Fuel Source & Equipment Type	Efficiency (%)	# Fuel Units Required	Fuel Unit
Natural gas (mid eff.) furnace	65	37	Cubic metres
Natural gas (high eff.) furnace	90	30	Cubic metres
Oil (mid eff.) furnace	65	39	Litres
Oil (high eff.) furnace	85	34	Litres
Electric furnace	100	29	kW·h
Air Source Heat pump	150	20	kW·h
Ground Source Heat pump	300	10	kW·h
Propane (mid eff.) furnace	65	64	Litres
Propane (high eff.) furnace	90	46	Litres
Woodstove	50-70	0.08 to 0.09	Cord (3.6 m³/128 ft³)

Heat pumps also operate in reverse as air-conditioners to cool homes, and they can supply a portion of hot water needs as well.

Heat pumps are different than other systems, in that they produce more energy than they consume. The way they are rated is through a 'coefficient of performance' (COP) or a Seasonal Heat Performance Factor (SHPF). Heat pump efficiency for cold climate operation has increased due to efficient, variable speed blowers, compressors and motors. Other improvements include larger coil surface areas; time delays on controls; and expansion valves to control the flow of the refrigerant more efficiently.

fuel sources

When you choose an energy source, you must strike a balance between the financial and environmental costs of the sourse as well as its renewability and availability. Electricity is 100 per cent efficient once it is in your house, but how the electricity is produced has an effect on how efficient it is overall. For example, electricity generated primarily by coal and oil-fired plants has energy penalties associated with it because the plants burn the fuel at 30 to 35 per cent efficiency. That means approximately 3kWh-equivalent (10,236Btu) of fuel are burnt to produce 1 kWh of electricity. In contrast, an oil furnace running at 80 per cent efficiency burns 1.25kWh (4,265Btu) to give you the same amount of space heating. In regions where hydropower is the main generation source, the associated greenhouse gases and waste energy are much lower. Like hydropower, wind, tidal energy, PV are all clean, renewable sources of energy. Nuclear energy has unanswered – and perhaps unanswerable – questions around safe operation and long-term storage of spent fuel rods.

When considering oil, gas or propane systems, look for high seasonal efficiencies. The overall efficiency may be high, but the important figure is the seasonal efficiency, which gives you an indication of the on-off cycling that the system goes through during the heating season. Any of these three heat sources can also be used to heat your water. Natural gas and propane are extremely efficient, and can fuel your cooking appliances and clothes dryer, as well as your space and water heating.

To avoid backdrafting, wood, gas, oil and propane heating units require a separate supply of outside combustion air. Backdrafting occurs when gases produced by burning fuel are drawn down the chimney due to negative pressure in the home. Energy efficient buildings are susceptible to back-drafting because they are more airtight than conventional buildings (see the section on indoor air quality for more on combustion air supply). A fresh air supply may be required near the heating source. Most building codes and regulations require separate chimney flues for combustion appliances.

Solid fuels – biomass and minerals – can be viable, inexpensive sources of backup heat for a solar home. Solid fuel can also be used in cookers, and in boilers that can provide heating and hot water for an entire house.

The most common biomass fuel used is wood. Wood is available in many forms including logs, manufactured logs, chips and pellets. If you are selecting a woodburning appliance in North America, look for an EPA-certified unit. Mineral fuels include bituminous coal, natural smokeless fuel (anthracite and dry steam coal), manufactured smokeless fuel and manufactured non-smokeless fuel.

Mineral and biomass solid fuels will usually result in higher local emissions than a high-efficiency natural gas-fired space heating system. Wood is often described as 'carbon neutral' – meaning that the amount of carbon dioxide emitted when the wood is burned matches the amount absorbed when the tree was living. However, this doesn't reflect emissions associated with forestry practices, transportation and manufacturing or processing of wood fuels.

Woodstoves are not recommended as a heating source in metropolitan areas because of the high costs of acquiring fuel and the possible environmental problems of burning large amounts of wood within a relatively small area.

In the UK, local authorities can declare the whole or part of their district to be a Smoke Control Zone under the Clean Air Act 1993. If you are in doubt as to whether your property is in a Smoke Controlled Area, contact the Environmental Health Department of your local authority.

Energy Source	Technology	Seasonal Efficiency (AFUE) %	Energy Savings % of Base**
Wood	Central furnace	45 to 55	
	Conventional stove (properly located)	55 to 70	
	High-tech stove (properly located)	70 to 80	
	Advanced combustion fireplace	50 to 70	
Oil	Cast-iron head burner (old furnace)	60	Base
	Flame-retention head replacement burner	70 to 78	14 to 23
	High-static replacement burner	74 to 82	19 to 27
	New standard furnace	78 to 86	23 to 30
	Mid-efficiency furnace	83 to 89	28 to 33
	Condensing furnace	85 to 95	29 to 37
	Integrated space/tap water mid-efficiency	83 to 89	28 to 33 space
			40 to 44 water
Natural Gas	Conventional	50	Base
	Vent damper with non-continuous pilot light	62 to 67	3 to 10
	Mid-efficiency	78 to 84	23 to 28
	High efficiency condensing furnace	89 to 96	33 to 38
	Integrated space/tap water condensing	89 to 96	33 to 38 space
			44 to 48 water
Electricity	Electric baseboards	100	
	Electric furnace or boiler	100	
	Air-source heat pump	1.7 COP*	
	Earth-energy system (ground-source heat pump)	2.6 COP*	
Propane	Conventional	62	Base
	Vent damper with non-continuous pilot light	64 to 69	3 to 10
	Mid-efficiency	79 to 85	21 to 27
	Condensing	87 to 94	29 to 34

Choosing the best energy source and heating system for your house is a juggling act between the fuels available to you, the equipment options in your region, and the impact of greenhouse gas emissions, based on the fuel used and the efficiency of your equipment.

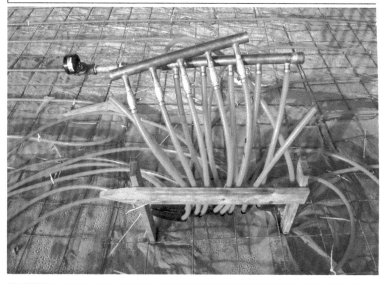

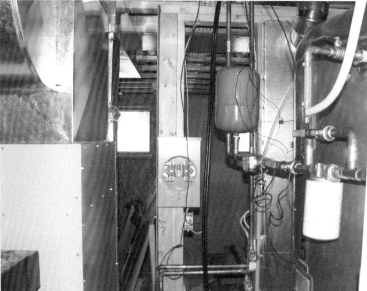

Heating oil produces 0.07311 kg of CO_2-equivalent per megajoule of energy used.
Natural Gas produces 0.04968 kg of CO_2-equivalent per megajoule of energy used.
To determine the greenhouse gas emissions from your oil or natural gas heating system, you need to multiply the figures above by the efficiency of the system (i.e., a 70 per cent efficient furnace produces 0.07311 kg * 1.30 (or 0.095 kg) CO_2-equivalent per megajoule of energy.

Opposite Page:
TYPICAL HEATING SYSTEM SEASONAL
EFFICIENCIES AND ENERGY SAVINGS
* COP = Coefficient of Performance, a measure of the heat delivered by a heat pump over the heating season per unit of electricity consumed.
** Base represents the energy consumed by a standard furnace.
Information from CANREN website www.canren.gc.ca/prod_serv/

INDOOR AIR QUALITY

Indoor air quality is influenced by many different factors. Lifestyles have changed, people spend much more time indoors than even a generation ago, and it is estimated that an average household has as many as 250 different chemicals stored inside. Often, the gases emitted by these chemicals are inert, but when placed beside another chemical in storage, two or more gases can combine and react, causing harmful vapours to be sent through the house. Many building materials contain chemicals which also 'outgas' when combined with household chemicals. In an airtight, energy efficient home with insufficient ventilation, these reactions can cause health problems. The focus of energy efficient buildings has been on how to keep the heated air inside the building. But without ventilation, there is nowhere for stale, polluted air to go. With rising numbers of reports of people being made sick by the buildings they live or work in, the need for increased fresh air supply and filtration has become apparent, but so has the need to reduce or eliminate pollutants at their source within the building envelope.

There are three key elements for a healthy indoor environment: ventilate the entire house, filter the air supply and eliminate sources of pollution.

DID YOU KNOW...
The World Health Organization estimates that 30 per cent of homes and buildings contain enough indoor pollutants to cause health problems for a significant portion of the population, ranging from sniffles to fatal allergic reactions. While building codes establish minimal ventilation rates and maximum available concentrations of proven pollutants, it is left to the homeowner to source and solve particular problems.

fresh air

Ventilation is needed to ensure good indoor air quality by exchanging stale indoor air with fresh outdoor air, and to provide additional air to combustion units such as stoves and furnaces. Until recently, ventilation was not a concern, because enough air leaked in through the building envelope to meet the needs of both the occupants and the heating system. This natural air exchange meant the buildings were drafty and required a lot of fuel for heat. New construction standards have made homes more energy efficient, in part, by greatly reducing the amount of air that leaks in and out of a home. The air quality in 'tight' buildings suffers, though, if there is poor ventilation. The resulting levels of humidity and various other pollutants can cause damage to the health of the occupants, as well as the structure of the home.

To compensate for the lost infiltration and exfiltration of air through cracks and holes in the building, we must install mechanical ventilation systems. Most local building regulations set out specific targets for mechanical ventilation in houses. Possible options range from high-capacity bathroom fans and vented range hoods to central fan systems and air-to-air heat exchangers.

In colder climates, energy savings warrant the installation of a heat recovery ventilator, which transfers most of the heat (up to 85 per cent) from the outgoing stale air to the incoming fresh air. Energy recovery ventilators (ERVs) are models that transfer moisture between the two air streams. ERVs are especially recommended in climates where cooling loads place strong demands on HVAC systems, or in areas with very cold winters, where extremely low indoor relative humidity levels can be a problem.

Relative humidity can be measured with a device called a hygrometer.

SOME SOURCES OF HUMIDITY

Like temperature, humidity stratifies (this is because warm air can hold more moisture). An air recirculation system can even out both temperature and humidity stratification.
Humidity is caused by breathing, showering, bathing, washing, stovetop boiling, hanging clothes to dry inside, and letting dishes air dry; by plants, aquariums, unvented clothes dryers, humidifiers and large quantities of wet wood (over 1/2 cord) stored inside. Humidity levels in a home are dependent on the lifestyle of the occupants, but as a rule, the more people in a home, the more moisture is generated.

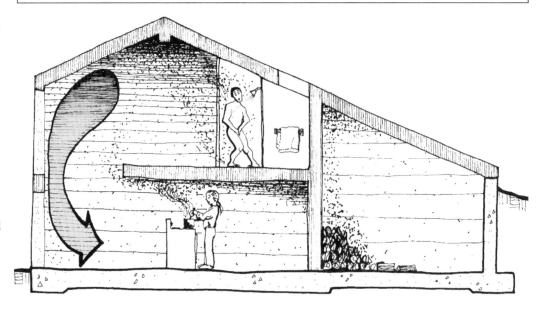

HUMIDITY

Maintaining even humidity levels is one of the main reasons airtight homes require a ventilation system. Humidity is the measure of the quantity of water vapour in the air. It is measured as a percentage of the 'saturated' state where air holds all the water vapour it can without some condensation occuring. As the capacity of air to hold moisture depends on its temperature, we call the quantity of water vapour (the level of saturation) in the air 'relative humidity'. For human comfort, we don't want the humidity too low. For the preservation of the structure, we don't want it too high. An average relative humidity level of between 40 and 60 per cent is considered healthy for both the inhabitants and the structure of a home.

High humidity levels can create problems with mold, mildew

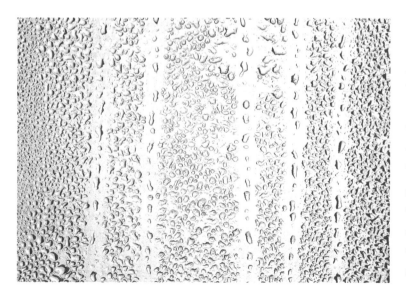

and rot. Condensation on cold surfaces and stains on finishes as well as unpleasant smells can occur in your home because of high humidity. Warm and humid conditions are good breeding grounds for bacteria. High humidity levels can also indicate high levels of indoor pollution, which can increase the occurrence of allergic reactions. On the other hand, low humidity can weaken wooden structural members and furniture by drying and cracking the wood. Static electricity is also a problem when air is too dry. Low humidity can cause sinus irritation, dehydration, dry skin, scratchy throats and contact lens discomfort.

You can reduce the amount of humidity in your home by removing the source of moisture and by reducing those activities or objects which generate moisture. Sources of humidity that cannot be removed (such as stoves, shower stalls and bathtubs) are best controlled by ventilation at the source: exhaust fans in each bathroom as well as range hoods. Clothes dryers that exhaust to the outdoors. Low humidity levels can be raised by bringing plants into the home or by inexpensive humidifiers.

Warmer air can hold more moisture. Assuming a constant level of water vapour in the air, the relative humidity can be lowered by heating the air. This is important to know when designing ventilation systems. Excess humidity is usually a problem encountered in the winter, when windows are kept closed. Airtight homes require fresh air intakes, which means cold air is coming into the home when the ventilation system is working the hardest. Preheating the moist fresh air before it enters the living space will reduce its relative humidity.

Humidity stratification occurs in your home in a similar manner to temperature stratification. To avoid this, an integrated air recirculating system can continuously mix the air mass, evening out the temperature and humidity levels throughout the building. Thermal mass can act as a 'flywheel' by absorbing humidity variations between seasons.

mechanical ventilation

A simple and inexpensive way to ventilate your home is to install exhaust fans and vents ('spot ventilation'), in those areas that produce moisture and odour: the kitchen, bathrooms, laundry (with dryer), and any other room planned for odour- or moisture-producing activities (such as painting or varnishing). Kitchen ranges and clothes dryers should have separate exterior vents, as grease-soaked lint is a definite fire hazard.

Adjustable barometric compensators allow fresh air inside the home in response to the negative pressure caused by exhaust fans or heating appliances. Manual and automatic wall intakes are also available. All fresh air intakes should be separated from exhaust outlets by a reasonable distance, and properly sized to counteract the negative pressure created by exhaust fans. Sizing and placement of fresh air intakes may be provided in your local building regulations.

SIZING SPOT VENTILATION

Kitchen range fans should have at least three speeds and move from 100 to 400 cubic feet per minute (cfm).

Bathroom fans should have at least 80 to 100 cfm capacity, and be wired to a timer or other control device.

Invest in high-quality, low-noise units to ensure that your household will use them. Look for bathroom fans with sone ratings no higher than 1.5. ('Sone' is a measure of loudness; one sone is about as loud as a common residential refrigerator.) Quiet kitchen range-hood fans are more difficult to find; often the best option is to use a remote, in-line fan.

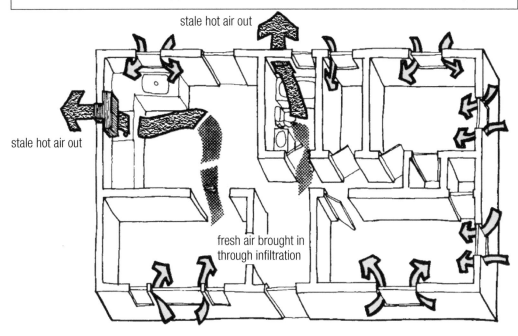

LOCAL FAN VENTILATION – BATHROOM & KITCHEN VENTED OUTDOORS

stale hot air out

stale hot air out

fresh air brought in through infiltration

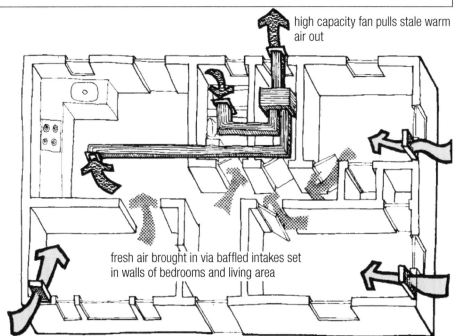

WHOLE HOUSE CENTRAL EXHAUST FAN

high capacity fan pulls stale warm air out

fresh air brought in via baffled intakes set in walls of bedrooms and living area

Central exhaust, or whole house ventilation systems can replace fans for bathrooms and a vented kitchen range hood. One high-capacity, low-noise fan is usually located in an attic, basement or utility room, and connected by ducting to several areas of the house. A multi-speed fan provides the flexibility for various operating speeds. Some manufactured central exhaust systems include individual room controls as well. Installing a dehumidistat (which will activate the fan when its sensors indicate high humidity levels) or a 24-hour programmable timer allows you to run the fan on low speed for a certain number of hours a day, and on high speed when moisture or pollutant levels increase.

Exhaust-only systems require fresh air intakes to replace the stale air that is removed. Pulling air out of buildings creates a negative pressure in the house, which the fan then has to work against, lowering its efficiency. Negative pressure could cause your chimney flue to backdraft, bringing combustion byproducts such as carbon monoxide into the living space – not very healthy. High levels of pollutants can also be drawn into the home from the building envelope itself (formaldehyde from some building products), or from the soil (radon and other soil gases).

Intakes can be connected to ducts that carry fresh air to several spaces, but the ducts must be insulated to prevent frost and condensation from forming on them. A heater installed in the duct will temper the cold air before it reaches the living space, thus avoiding uncomfortable drafts. Heating the fresh air will also reduce its relative humidity. A cost-effective way of ensuring adequate fresh air during the winter months is to connect a fresh air intake duct to the return side of your furnace. This allows fresh air to be drawn into the furnace plenum, and so heats the fresh air before it is distributed to living areas.

Heat recovery ventilators (HRVs), or air-to-air heat exchangers, and energy recovery ventilators (ERVs) are known as 'balanced pressure ventilation systems'. They have two fans – one to

exhaust air and one to pull air in from outside. Heat is extracted from the outgoing exhaust and returned to the home by heating the fresh air. Up to 85 per cent of the heat can be 'recycled' when the unit is properly installed and balanced. The unit must be sized correctly for your home, and the fresh air/exhaust air fans must be balanced. To optimize the efficiency of the unit, all ductwork inside the house should be smooth sheet metal, and the duct runs should be as short as possible.

SCHEMATIC OF A PLATE-TYPE HRV

The most common type of HRV contains a plate-type core which allows heat transfer but prevents the incoming and outgoing streams of air from mixing and causing contamination of the incoming air.

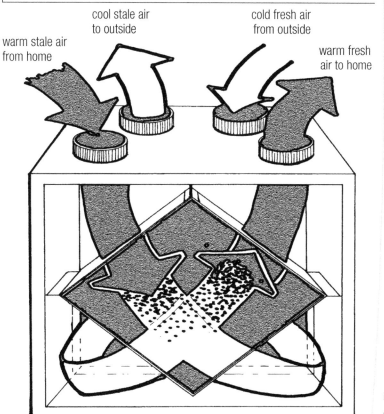

cool stale air to outside

cold fresh air from outside

warm stale air from home

warm fresh air to home

All ductwork should be run through the conditioned space of your house. Otherwise, you are courting disaster in the form of frost build-up and mold growth.

The heat exchanger is the heart of an HRV, usually consisting of a cube-shaped transfer unit made from special conductive materials. Incoming and outgoing airflows pass through different sides of the cube (but are not mixed), allowing conditioned exhaust air to raise or lower the temperature of incoming fresh air. ERVs also allow the exchange of moisture to control humidity. This can be especially valuable in situations where problems may be created by extreme differences in interior and exterior moisture levels. For instance in cold, heating-dominated climates, better air flow and the controlled introduction of humidity to a dry indoor environment can help control wintertime window condensation. In humid summer climates that are cooling dominated, it can be critical to dry out incoming air so that mildew or mold do not develop in ductwork.

Some HRV and ERV models have a built-in 'recirculation' mode, which allows you to ventilate the whole house 20 minutes out of each hour, and recirculate the house air for the other 40 minutes. This option offers a cost-effective way to ventilate and destratify a passive solar home that does not use a forced-air system to distribute the solar-heated air.

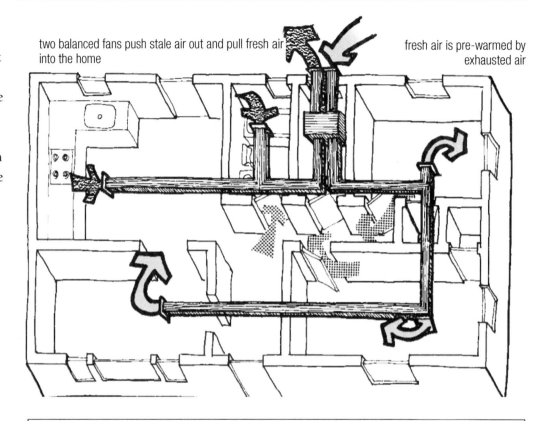

two balanced fans push stale air out and pull fresh air into the home

fresh air is pre-warmed by exhausted air

VENTILATION COMPARISON TABLE

System	Advantages	Disadvantages
local exhaust fans	low initial investment removes local pollutants effectively multiple controls for flexibility	noise levels higher operating cost includes heating fresh air creates air leaks & drafts potential for back drafting
exhaust only (central fan)	average initial investment removes local pollutants effectively remote fan is quieter suitable for constant operation	more complicated installation higher operating cost includes heating fresh air potential for back drafting potential de-pressurization problems in airtight construction
heat recovery ventilator (HRV)	recovers up to 85 per cent of heat from exhaust air minimizes air leaks & drafts lower operating cost when properly balanced & sized pays for itself in 4-12 years	higher initial investment complicated installation potential for over ventilation can lead to higher energy costs if not properly sized & balanced

RECIRCULATION

Recirculation is the constant movement and mixing of the total air mass within the living space. This minimizes temperature and humidity stratification, diffuses pollutants into the total air mass, and allows for treatment of the air by filtration, humidification or de-humidification, and the addition or removal of heat from the home. Ideally, the ventilation and air recirculating systems should be integrated. A fully integrated system provides air circulation, auxiliary heat, controls temperature and humidity stratification, and air quality. In a house that is extremely well-insulated, airtight, with good solar aspects, an integrated mechanical system can take the form of a small forced air furnace, or an HRV with a heating coil installed in the house air supply duct.

COMBUSTION AIR

For airtight homes, most building codes require an adequate supply of fresh air for occupants, and a separate supply of fresh air for all fuel burning appliances requiring combustion and/or dilution air. The reason for this is fuel-burning appliances create negative pressure inside the home. This can result in backdrafting, which can lead to carbon monoxide poisoning. 'Fuel-burning appliances' can be defined as fireplaces, oil and propane-fired furnaces and boilers, and all wood-fired appliances including stoves, furnaces and boilers.

There are several options for supplying fresh air to oil and gas furnaces and boilers: purchase a unit with a 'sealed' combustion chamber where the fresh air is ducted directly to the burner; purchase a unit which is equipped with connections to accept outside air directly; or install an outside air duct with a motorized damper wired to the burner control. In some cases a simple outside air duct through the wall near the appliance may be acceptable, but it is not recommended.

Open fireplaces are not recommended in airtight homes. Advanced fireplaces, which have gasketed doors to decrease air leakage problems, have little or no interaction with the house air, so the chances of combustion products spilling into the home are minimal. In addition, they greatly enhance the effectiveness of the fireplace.

SYSTEM CONTROLS

manual switch

crank timer

humidistat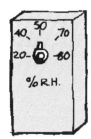

variable speed

24-hour timer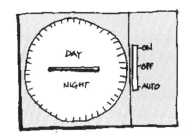

Many wood stoves and furnaces connect directly to a fresh air feed. In some provinces, this outside air feed is mandatory for any wood-burning appliance. Install a positively-closing manually operated damper to keep air out when the unit is not in use.

Ventilation systems can be controlled by humidistats, switches and wind-up timers, but, for the system to be effective, you have to use the controls! Programmable timers are a boon to busy households, as they turn on the system when required, and continue to operate without further attention. Computerized 'smart' ventilation systems are also available.

natural ventilation

Incorporate as much 'passive' ventilation as possible. Avoiding energy-intensive air-conditioning systems (some of which do not supply fresh air, but only recirculate stale indoor air) in the summer. Passive ventilation relies on the 'stack effect'. The natural forces of wind and temperature difference between indoor and outdoor air draw fresh air through the home. The temperature difference between warm indoor air and cooler outdoor air causes the stack effect. Warm air rises out through high openings while simultaneously drawing in cooler, fresh outdoor air through low openings.

Passive ventilation relies on natural wind forces to be effective. This means you must know the direction of the prevailing summer winds, daily variations and possible blocking by buildings, trees and hills. Passive ventilation also relies on the pressure systems in and around your home. The exterior wall that faces the prevailing wind is a high pressure zone, while the roof and sides away from the wind are low pressure zones. High pressure zones push air into the home, while low pressure zones react with the pressure inside the home to pull air out of the home.

For summer cooling, lower windows which face the prevailing breezes will act as inlets, while upper windows on the opposite side, or in the low pressure areas of the roof and sides of the building will act as outlets, drawing the cooler air through the building. Keep the windows closed during the day to keep the heat out. Heat collected during the day will be absorbed by the thermal mass. Open windows or vents at night to ventilate and cool the interior thermal mass. Un-

less you have oversized the amount of south-facing glazing and undersized the amount of thermal mass in your solar home, you should experience little over-heating.

SHADING
Shading direct gain areas to the south will be a key factor in keeping your house cool during the summer months. Over-hangs can be used to block the summer sun but allow winter solar gain. Overhangs

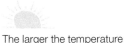

The larger the temperature difference between indoor and outdoor air, the greater the height between inlets and outlets, and the larger the openings, the greater the flow of air. Your inlets and outlets can be operable windows, or manual or automatic vents.

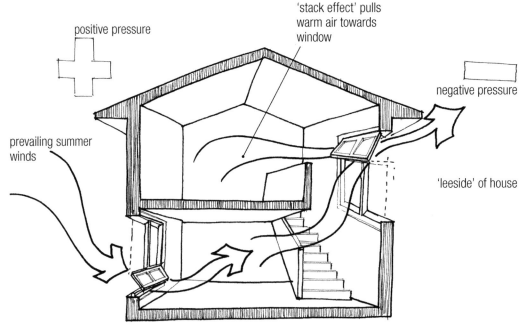

positive pressure

'stack effect' pulls warm air towards window

negative pressure

prevailing summer winds

'leeside' of house

don't have to take over the south facade of your house. When you are outside at noon in the summer, the sun is high overhead, and you will cast a short shadow on the ground. However, the short brim of your baseball cap casts a long shadow across your body. Likewise, the higher, or more vertical the arc of the sun, the longer the shadow that an overhang generates along the face of the wall. Summer shadows extend down walls the furthest, winter shadows the least. The further you move from the equator, the shorter the wall shadow, assuming the overhang length. So, the latitude of your site, combined with the width of the overhang and it's height above the 'head' of the window will dictate how much shading actually occurs during the summer when you need it. There is as yet no universally simple formula for sizing overhangs. While one overhang methodology works well for some locations, it can be completely inappropriate for others. That being said, the following rule of thumb will give you a good rough cut on overhangs.

Cold climates: locate shadow line at mid-window using the June 21 (summer solstice) sun angle.

Moderate climates: locate shadow line at the windowsill using the June 21 (summer solstice) sun angle.

The long, hot afternoons and evenings translate into massive heat gain through windows facing 30° either side of west. To be of any value overhangs must be in the form of covered porches. However, they may block winter afternoon sun. Trellises, arbours, 'brise soleils' and retractable awnings can be used. Seasonal shade cloths can also be used, and are very popular in hot countries like Australia. Fast-growing perennial vines that die back in the fall, such as hops (10m or 33ft a year) and some types of clematis (3m or 10ft a year or more) create beautiful, cool outdoor rooms that still allow the low-slung, short-lived afternoon winter sun into the house.

Use prudence when planting grapes, kiwis and other food crops for shade – especially around doors and over sitting areas, as they will make a sticky mess that will be tracked inside and will attract wasps as well as bigger pests.

Photo Courtesy of Chris Mattock, Habitat Design + Consulting

The house to the left has an overhang to the south, putting the south-facing glazing on the main floor in complete shade during the summer. Also, the overhang to the west is quite broad, reducing the chance of overheating in the afternoons.

The optimum overhang design for your house may not mesh with local zoning ordinances. If your house is on or close to the required setback, you can move the house further back or redesign your windows or your overhang, or apply for a variance (an exception to the ordinance).

pollutants

Pollutants can include particulate matter (dust, pollens, fibers, soot and animal danders); gases (water vapour, cooking odours, radon and outgassing from manufactured products), and biological matter (bacteria, viruses and molds).

Pollutants can be controlled in three ways: source removal, dilution and filtration. The first line of defense is to remove or seal off pollutant sources. Your second line of defense is dilution, or removal at source through ventilation. Windows, vents and ducts require no extra mechanical apparatus. Extractor fans and whole-house ventilation systems are more complex. Your third defense is to treat the air by filtration. Use electrostatic, charcoal, or fibre filters, as well as dehumidifiers or humidifiers if necessary. Whole-house ventilation and a filtered air recirculation system can eliminate many of the pollutants that make us ill. However, if filters, humidifiers or dehumidifiers are installed, you must maintain them well so they do not become breeding grounds for bacteria that could spread throughout the home.

DUST AND PARTICULATE MATTER
Many people suffer allergies from dust and other airborne particulates such as pollen and animal dander. One way to reduce the amount of irritating particulates in your home is to minimize upholstered and carpeted surfaces, as they tend to harbour the stuff.

Another way of reducing dust and particulate levels is to invest in a central vacuum system that exhausts to the outside. Most canister-type vacuum cleaners filter out the bigger particles, but spew the finer, most irritating bits right back into your house. Although the initial investment may be higher than a canister vacuum, the long-term health benefits can outweigh the original cost.

CO AND CO_2 EMISSIONS
Carbon monoxide (CO) and carbon dioxide (CO_2) are both colourless, odourless gases. CO is the result of incomplete combustion. The more dangerous of the two, it is readily absorbed in the bloodstream, depleting the ability of blood to carry oxygen.

Sources of CO include poorly vented combustion appliances as well as automobile exhaust drawn in from an attached garage or busy street. For safety, CO detectors should be hard-wired in rooms with wood stoves and other combustion appliances. In some regions it is mandatory to have CO detectors installed. The most important step you can take is to ensure that CO never has an opportunity to enter your home.

CO_2 is normally present in air in small quantities, usually around unvented combustion appliances, or in areas with poor fresh air supply. CO_2 is also a by-product of human respiration. Increased levels are perceived as 'stale air' and 'stuffiness', and easily sensed before becoming dangerous. Plants can be used as a control, as they take CO_2 in and give off oxygen. Other controls include CO_2 detectors that can be wired into your ventilation system to bring in more fresh air when CO_2 levels increase.

Indoor air quality is measured by pollutant type and concentration. Concentrations are affected by the entry rate, release rate and generation rate of the sources. These, in turn, are affected by the volume of air in the house, the rate of air movement and the rate of air changes per hour and whether or not the air is being filtered.

VOLATILE ORGANIC COMPOUNDS

Probably the most common pollutants in buildings today are volatile organic compounds (VOCs), chemicals that readily evaporate and irritate the eyes, nose and throat. They are found in many different building materials, from carpets to paint. Carpets can be one of the main culprits, as the plastics and chemicals used to resist stains, bond fibers and hold colours all emit VOCs, as do the fungicides, pesticides and bactericides with which carpets are routinely treated.

Wood floors, tile floors and natural fibre area rugs are viable alternatives to synthetic wall-to-wall carpeting, as you can easily wash and clean both the floor and rugs. As well as cutting down on the VOCs, hard-surfaced floors, with or without area rugs, drastically reduce the amount of irritating dust and microorganisms in the air.

If wall-to-wall carpets are a must, lessen their impact on your health. Allow new carpets to outgas for a few days by unrolling them in a covered storage space. When installing them, don't use glue; tack them into place. Underlays should be tacked-down jute or other material that will not outgas. Another way to reduce VOCs emitted by carpets is to 'bake' them out. The irony of carpetting is this: at the same rate new carpet outgases, it becomes a source of dust and other particulate matter. Your old carpet may be reasonably inert in terms of VOCs, but every time you walk across it you create little whirlwinds of dust and dander.

Vinyl wall coverings, wallpaper glues and upholstery fabrics may also contain high levels of VOCs, especially synthetic fabrics, although natural fabrics may also contain irritating chemicals. Look for untreated natural fibre fabrics and low-toxicity wallpaper glues.

Use bactericide and fungicide-free paints, stains and varnishes. No and low VOC paints are readily available from major paint manufacturers. Water-based latex paints, sealants and finishes, contain a fraction of the solvents used in petroleum-based products. Avoid wood preservatives or treated wood products, especially indoors. Allow plenty of time (several days) for paints, stains and varnishes to dry before inhabiting the space; use a large fan for ventilation.

FORMALDEHYDE

Formaldehyde is still a big player in air quality problems. Millions of pounds of it are used as adhesives in building products such as plywood and particleboard. With so much of today's furniture and cabinetry made from these materials, it's hard to avoid. Use exterior grade plywood as it contains phenol formaldehyde resin, which is more stable than the urea formaldehyde used in interior grade plywood. Apply a water-based no VOC sealant as a vapour barrier on all accessible areas of particleboard furniture. Choose furniture made from solid wood or other natural materials such as rattan or wicker.

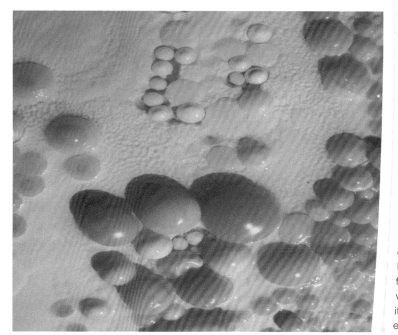

Formaldehyde, a VOC, made news several years ago in the form of urea formaldehyde foam insulation (UFFI). After many people complained of health problems because of outgassing from the foam, the product was taken off the market, and it was stripped out of many existing homes.

RADON

Radon is a naturally occurring gas caused by the decay of radium in the soil and accounts for up to 50 per cent of all the radiation we receive from natural sources. This tasteless, odourless gas occurs throughout the world and has been linked to lung cancer. Moisture in the soil traps the radon so the gas accumulates under the foundation of a house. When the moisture evaporates, the soil releases the gas in heavier concentrations. The higher the moisture content in the ground, the greater the subsequent emission of radon gas.

The amount of radon in a home depends on such factors as soil characteristics, the condition and construction of the foundation, the rate of air exchange, the lifestyle of the occupants and the weather. Because of these varying factors, the only way to determine radon levels in your home is to test for it.

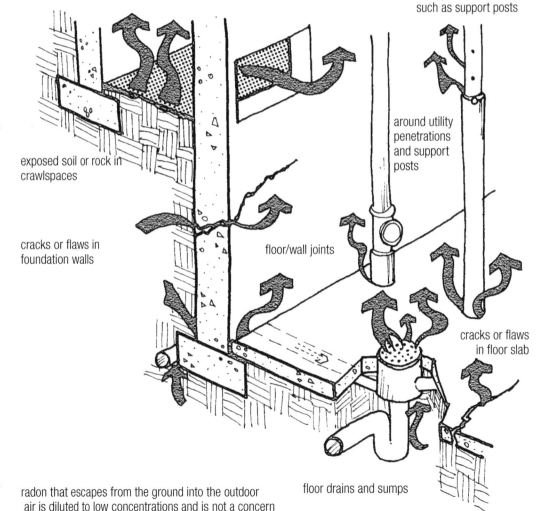

HOW RADON GETS INTO YOUR HOME

The air pressure inside a house is usually slightly lower than that in the soil that surrounds the foundation. This difference in pressure causes the air and other gases in the soil, including radon, to be drawn into the home.

through hollow objects such as support posts

around utility penetrations and support posts

exposed soil or rock in crawlspaces

cracks or flaws in foundation walls

floor/wall joints

cracks or flaws in floor slab

radon that escapes from the ground into the outdoor air is diluted to low concentrations and is not a concern

floor drains and sumps

The list of chemicals and materials affecting indoor air quality is only a partial one. As research goes on in the field of healthy housing, more and more questions come up as to the safety and health problems associated with every and any building material. There are several books and online resources available that are dedicated to these issues.

This is a tricky area, because there are so many variables, and often people don't know that they, or a family member, are sensitive to a specific chemical or combination of chemicals. By avoiding or removing the sources of the big health hazards – formaldehyde, VOCs and radon – and by designing in an efficient ventilation and air filtration system, you can put yourself and your household in a reasonably healthy environment. How far you take it depends on your sensitivities and your budget – some of the alternatives are not cheap. But don't put yourself in the position of compromising your health because of the up-front costs of materials.

POLLUTANTS AND WAYS TO CONTROL THEM

pollutant	source	source control	Dilution control
CO, CO$_2$	combustion by-products from gas or wood burning appliances	oil/gas furnaces with sealed combustion chambers, pilot free units	fresh air feed to combustion unit
RADON GAS	earth, stone, granite, (infiltration through foundation), phosphogypsum products (plaster, cement & plasterboard)	seal foundation cracks, use natural gypsum plasterboard or lime plaster	increase ventilation, use radon sump
VOCs	paints, varnishes, stains, adhesives, synthetic carpets & underlays	use non-chemical glues, low solvent paints, natural fiber carpets & jute underlays, area rugs	
FORMALDEHYDE	particle board & plywood adhesives; synthetic fabrics used in carpeting, underlays & upholstery; vinyl plastics in tiles, electrical equipment, imitation wood panelling & wallpapers; oven & carpet cleaners	avoid particle board & plywood construction materials & furniture; use solid woods, 'low-emission' formaldehyde boards, natural fabrics & carpeting, alternative cleaning products	
HEAVY METALS	pipes & water supply (lead, cadmium, mercury, aluminum, iron, magnesium, copper), paints (lead & cadmium), cookware (aluminum)	replace lead & lead soldered pipes, test water, install filtration systems, use low-toxin paints, stainless steel, glass or enamel cookware	
HAZARDOUS CHEMICAL TREATMENTS	timber preservatives (lindane, pentachlorophenol (PCP), etc.); pesticides & fungicides	avoid preserved lumber, practise biological pest control	
DUST & OTHER BIOLOGICAL POLLUTANTS (airborne fungi, micro-organisms, bacteria, etc.)	many sources, including carpets, upholstery, draperies, animal dander, excessive humidity levels	avoid wall-to-wall carpeting, use ceramic tiled floors with area rugs, install central vacuum system to exhaust dust directly outdoors, keep pets outdoors	ventilate to outdoors install & maintain air filtering system

CMHC has several free publications on indoor air quality and housing for the environmentally hypersensitive

THE BUILDING ENVELOPE

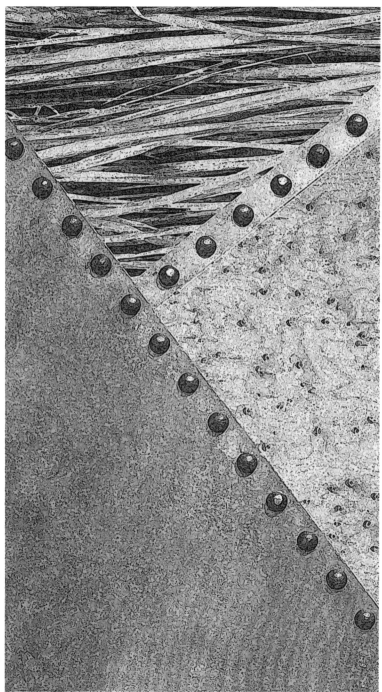

The exterior surfaces of your house protect you and the interior of the building from the elements. The building envelope is a major portion of the investment in your house, and as such, should receive a large amount of your attention. The choices you make when building the envelope last for a very long time. Every decision affects the performance of the house; the amount of solar gain that can be collected and stored, the overall energy requirements of the house and the durability of your investment in the house. 'Fixing' the envelope after construction is a costly endeavour and best avoided.

Think this through: instead of putting short-lived equipment and finishes (such as appliances and carpeting) on your mortgage, focus your long-term debt on long-term investment in the envelope of your house. Your energy bills will be lower, allowing you to put some of your monthly income aside for appliances, carpeting and so on. Your overall resale value will improve, as well.

An energy efficient house starts with a well-sealed envelope and adequate ventilation capacity. Insulation is added to floors, walls and ceilings. High-performance windows wrap up the job. There are many ways to accomplish these steps, and determining the right combination of materials and methods can be confusing. Different messages from various building groups and organizations don't make it easier. As well, you have to take into consideration climatic requirements as well as regional building styles.

New construction or retrofit projects can benefit from a home energy audit. The EnerGuide for Houses rating system places your home on a scale of 0 to 100 when compared to houses of similar style and vintage. Having a rating on your house will serve to show prospective buyers how well your house performs. While currently voluntary in Canada this type of rating system has become mandatory in several European countries when selling a house.

airtight envelope

Typically, for at least six months of every year, houses in cool climates lose huge amounts of heat due to air leakage. That's a lot of wasted energy! There are a few schools of thought on this topic. We have chosen to go with the approach taken by most researchers and building experts in cool climates: start with an airtight envelope and add controlled mechanical ventilation to maintain good indoor air quality.

Most government agencies have publications available that deal with air sealing or weatherization in cool climates. In the US, the Department of Energy's Energy Efficiency and Renewable Energy (EERE) Information Centre is a good place to find information and resources. Under its Heat and Energy Saving System, the UK aims to reduce every building's carbon emissions. Search using keywords: weatherization, air sealing, residential

AIR TIGHTNESS TESTING

Energy savings from air sealing alone can be significant – up to 35 per cent in newer homes, and even up to 50 per cent in older homes. Most of the work can be done at little cost. But how do you know if you've been successful? There is a standard air leakage rate test, carried out using a 'blower door' apparatus. A large, adjustable fan sits in a frame that fits snugly into an exterior door. The rate at which air flows through the fan (equal to the air leaking through the building envelope) and the pressure difference across the building envelope determine the air leakage rate, measured in air changes per hour (AC/hr) – how many times the volume of air in the house is exchanged in one hour under test conditions. In some regions, the AC/hr is calculated from a depressurization rate of 50 Pascals, in others, 10 Pascals is used. Some energy efficiency programmes and countries that require blower door tests: Energy Star (US), all new buildings in Ireland and the UK. The PassivHaus standard requires that a house meet a specific target for air leakage, as does the R-2000 programme in Canada and Japan.

Comparison of Air Change Rates (AChr@ 50Pa)	
Standard Current Construction	3.5 - 5.5
R-2000	1.5 or less
PassivHaus	0.6 or less

New construction offers lots of strategies for improving air tightness. Retrofits and renovations are trickier. Your air tightness strategies need to consider the movement of water vapour. Moisture transfer by air currents is very fast. Moisture within wall cavities causes damage to the envelope. The section on climate control demonstrates the dynamic relationship between heat, air and moisture.

A quick recap: water vapour moves in and out of a building in three ways: with air currents, by diffusion through materials and by heat transfer. Of these three, air movement is the crucial force to deal with. This is because air moves from a high pressure area to a lower one by the easiest path possible: through any available hole in the building envelope. Moisture transfer by air currents accounts for more than 98 per cent of all water vapor movement in building cavities.

VENTILATION REQUIREMENTS

When a house is airtight, it is imperative to compensate for the reduced amount of air leakage by introducing controlled mechanical ventilation.

WHAT TO DO AND HOW TO DO IT

Air barriers and vapour diffusion retarders (VDRs) are integral parts of the building envelope. Here is a quick overview of the difference between air barriers and vapour diffusion retarders.

Air barriers block air movement through building cavities. A continuous air barrier not only saves energy by reducing air leakage and heat loss, but also stops water vapour carried by the air. There are two types of air barriers: membrane (polyethylene, foil or specifically formulated paints) and rigid (plywood, drywall or appropriately rated rigid board insulation). Both are effective when installed properly. Many of these materials are also used for other purposes in the house, so they do double duty. What to choose and how to use it depends on your region's climate.

A vapour diffusion retarder (VDR) slows the diffusion of water vapour and so helps prevent moisture problems. VDRs are classified according to 'perms', with lower ratings, indicating more effective retardation of vapour diffusion. The VDR must be located on the warm side of the insulation to prevent condensation in the wall. As a rule of thumb, in most cool climates this means two-thirds of the insulation has to be between the exterior and the VDR. In more northerly areas, this might need to be increased to four-fifths.

What you use for insulation influences what you use for an air barrier and VDR. Fibrous insulations don't stop air movement, so they need a continuous film, such as a 6ml polyethylene sheet installed on the interior side of the wall. This type of air barrier also acts as a VDR as long as it is completely continuous. On the other hand, closed-cell spray foam insulations, act as insulation, air barrier and VDR all in one application.

How the house is put together also determines what you use for an air barrier and VDR. Because they need to be truly continuous, membrane air barriers, such as polyethylene, can really only be installed properly during the initial construction phase of a new house, or as part of a 'gut the entire house' renovation. A rigid air barrier method can work well in existing houses. Using the 'airtight drywall approach', the drywall is sealed to the underlying framing and subfloors, as well as at all junctions and openings using gaskets and caulkings.

One of the benefits of this approach is that the air barrier is visible to the interior, so it can be repaired after construction. To transform the drywall into a VDR, low-permeance interior paint is applied.

SEALED WITH... AN EXTERIOR AIR BARRIER?

An exterior air barrier must allow adequate vapour transmission while stopping air movement, or be located so that it prevents condensation within the wall. One of the best opportunities to really improve an existing building envelope is when you replace the siding on your house. You can apply an exterior air barrier and increase your insulation levels without the labour required to seal all those nooks and crannies on interior walls and ceilings. Once the walls are stripped of the old siding, a contractor can fill any empty wall cavities with blown-in insulation. Then, the exterior of the house can be covered with layers of rigid board insulation. Taped and sealed at the seams and edges, it forms a continuous exterior air barrier, doing two jobs at once. Another common material used to create an exterior air barrier is fiberous spun polyolefin plastic 'house wrap.' Some wraps have better weathering or water repelling abilities than others. House wraps are usually wrapped around the exterior of a house during construction. Sealing the joints improves the wrap's performance by about 20 per cent.

Where the same material is used for the air barrier and the VDR, it must be continuous. If you install a separate, continuous air barrier, then it is not so critical that the VDR be continuous, although it is still very important.

In new or existing projects, more than one approach can be used to air seal your house. A 'hybrid' air barrier can be made with drywall sealed to an exterior air barrier material carried around the floor assembly to join the rigid air barrier below or above.

insulation levels

Insulation is commonplace in cool climate homes, whether new or old. But how much is enough? To what lengths should you go to reduce your energy needs through improving your thermal envelope?

Most energy efficiency programmes have suggested ranges of insulation for above grade walls, ceilings, foundations and exposed floors. Beyond these levels, there are diminishing returns. It is true that you get the biggest bang for the first few centimetres or inches of insulation and after that, the amount of energy you save per unit of thickness goes down exponentially. However, a house is a system, and no one component (i.e. insulation) acts in isolation from another (i.e. space heating system).

Several key energy efficiency standards and programmes were created more than 20 years ago in response to high fuel prices and low insulation levels. These standards were also designed to integrate into the building industry without too much disruption in the way builders put houses together – or too much increase in the up-front capital cost.

Most energy efficiency programmes for existing houses look at ways to reduce energy use in space and water heating by 20 to 25 per cent. With ever-increasing costs for heating, the recommended energy efficiency measures can create 'lost opportunities' for deeper energy reductions.

Twenty years ago, windows were the bane of any energy efficient house, especially a passive solar house with lots of glazing. It didn't matter how much more insulation you put in walls and ceilings – single-pane windows in wood frames basically cancelled it out. Even today, windows are still the weakest point in the building envelope.

All these factors have driven what the industry considers to be 'normal' insulation levels. But here's the thing: your assumptions make a difference. If the underlying assumption is that the house require a typical heating system, then the house is built to accommodate that heating system and use it efficiently. If the underlying assumption is that there will be only a backup heat source, then the house is built to collect and conserve energy. Strategies change as assumptions change.

New window technology, new insulation materials, better understanding of building science and the prospect of ever-increasing fuel costs have re-sparked great interest in 'superinsulated' houses. The history of superinsulated houses, starts back in the 1970s with Wayne Schick at the University of Illinois, and the 'Lo-Cal' housing concept, in Canada, with Harold Orr and the Saskatchewan Research

Council and the 'Saskatchewan House' and in Massachusetts with the 'Leger' House. Current programmes and standards that focus on highly insulated envelopes include Low Energy Houses (US), the PassivHaus (Passive House) (EU), the Equilibrium Initiative (Canada) and the EcoHouse (UK).

The principles of super insulation were developed to make wood-framed houses more thermally efficient. Wood framed houses are considered lightweight construction – rapidly heating up and cooling down on a daily basis. Superinsulation was developed to slow down the transfer of heat and make wood frame houses more thermally efficient. It has been common practice to construct thicker walls so that more insulation could be installed. For very thick walls, double stud may be more economical than a composite section. In the past, the number of widely-available insulation materials was limited. Today there is a good selection of insulation types to choose from and some, particularly rigid urethane and rigid foam boards with reflective facings, achieve very good thermal values within narrower wall cavities. As well, some products act as air and moisture barriers, eliminating materials and labour, as well as improving the envelope.

SUPER INSULATED HOUSE DESIGN
TYPICALLY INCLUDES:
High value thermal insulation in the roof, walls and ground floor
Continuously insulated building envelope with no cold bridges
Airtight construction, especially at doors and windows
Triple glazed windows
A mechanical heat recovery ventilation system
A minimal back-up heating system
Triple glazing

Specially targetted areas for spray foam applications include rim joist/header areas, wall/ceiling/floor junctions and odd framing areas such as dormers and bays as well as unvented roofs.

fibrous insulation

Fibrous insulations are most commonly used inside wall cavities and attics, in either batt or blown-in form. There are several types of fibrous insulation, ranging from the ubiquitous fibreglass to mineral wool and cellulose. Insulation is also made from waste agricultural fibres such as cotton, sheeps wool, hemp and straw.

FIBREGLASS
Fibreglass accounts for most of the insulation in homes today, as it is relatively inexpensive and easily installed. Higher density batts, with about 15 per cent more insulating ability than standard batts, increase the overall insulation level because the thinner batts fit into more angles and smaller spaces with less compression. Fibreglass insulation has up to 40 per cent recycled content.

MINERAL WOOL
Made from recycled slag and mined basalt rock, mineral wool is naturally resistant to fire and pests and is highly sound absorbent. Mineral wool is available in batts (for wall and ceiling cavities) or as semi-rigid boards (for exterior below-grade use). Made with at least 40 per cent recycled material, mineral wool is held together with a thermosetting binder (often a food-grade starch).

CELLULOSE
Made from shredded newsprint containing up to 85 per cent recycled material and 15 per cent borate-based fire retardant, blown-in or sprayed-on cellulose has a higher insulating value than fiberglass. A well-done, wet-blown dense-pack installation can reduce air leakage and get into hard-to-reach areas. Wet-blown installations may have difficulty drying in certain climates.

foam insulation

Primarily there are three classes of petroleum-based foam insulations: polystyrene, polyurethane and polyicynene. Most types of foam insulation are available as spray-on foams or as board stock. Spray-on foams are very useful in new and retrofit projects. Spay-on foams are also very effective when insulating hard-to-reach areas. Foam board-type insulations are commonly used as insulating sheathing.

Foam insulations have a wide range of insulating values, based on whether they are low-density 'open cell' foams or high-density 'closed cell' products. Low density, open cell products offer high acoustic reduction, provide a good air barrier, but act as a moderate vapour diffusion retarder. They are less expensive due to their lighter densities. Closed cell, high-density products are better insulators and can function as both an air barrier and vapour diffusion retarder. As well, their greater compression strength allows them to be used in a wider variety of applications.

Ozone-depleting blowing agents such as CFCs and HCFCs are a cause for concern when specifying foam insulations. The major problem with both compounds is chlorine content. When CFCs or HCFCs enter the stratosphere, radiation in the atmosphere breaks down the molecules, releasing chlorine, which destroys ozone at a rate of 100 ozone molecules per 1 chlorine molecule.

SPRAY-ON FOAMS
There are several varieties of spray-on foams. They start out in liquid form, then expand and solidify almost instantaneously filling minute cavities, cracks and crevices. Spray-on foams usually cost three to four times more than traditional fibreglass insulation and must be installed by professionals.

Use light-gauge steel studs for interior partition walls. Before building partition walls, apply vapour barrier and drywall to the entire surface of exterior walls. This costs about the same as wood (about 10 per cent higher material cost but generally lower labour cost). Steel studs also have the benefit of transmitting less sound than a 2x4 wall and increasing fire resistance of the structure. And since they don't warp you won't get nail pops.

Ralph Doncaster

board insulation

Polyurethane foam is a closed cell foam, available in four densities, each suited to a set of applications. Higher density, means higher insulating value and lower vapour permeability. Higher density also means higher durability. Most polyurethanes use ozone-depleting blowing agents.

Polyicynene is a low-density, open-cell spray-on foam insulation. It is an environmentally safe product that does not off-gas and contains no formaldehyde, CFCs or HCFCs. In addition to effective thermal performance, it is effective in air-sealing the building envelope.

Cementitious foam insulation is made from magnesium oxide derived from sea water. It is blown in place with air. No CFCs or HCFCs are used. Because of its inorganic composition, it has very low VOC emissions, is totally inert and non-combustible. From the standpoint of indoor air quality, cementitious foam is also more costly. This type of insulation is considered the most benign. Most often used in commercial applications, but the material is adaptable to most wall, roof and ceiling cavities.

There are three types of rigid foam insulations: molded expanded polystyrene (MEPS), extruded expanded polystyrene (XEPS) and urethane (polyurethane, polyisocyanurate, etc.), presented in order of increasing insulation value. All three foams are combustible and must be covered with a fire-protective finish on any exposed interior surface.

Low-density MEPS (Type I) are suitable for sheathing above grade exterior walls, interior basement walls, above grade interior walls, flat roofs and cathedral ceilings. High-density MEPS (Type II) can be used to insulate exterior below-grade walls. Extruded expanded polystyrene (XEPS) is a 'closed cell' foam, available in a range of densities (Type II, III, and IV). The higher the density (indicated by the type number), the higher the insulating value.

Polyurethane, polyisocyanurate and phenolic above grade board foams are closed-cell structures that can be used as interior or exterior wall sheathing and for interior below grade applications. These are not suitable for exterior below grade applications.

Photo courtesy of Crowell Construction

Photo courtesy of Crowell Construction

You can get a small (but really cheap) boost to thermal mass by being selective about drywall thicknesses. A 5/8' thick sheet increases the drywall mass by 35 per cent (from 1.6lbs/sf to 2.2 lbs/sf) over a standard 1/2' sheet at cost increase of less than 10 per cent. There's the added plus of significantly increased fire resistance. Downside is extra work for finish carpenters to furr out pre-hung door jambs an extra 1/4', and drywall hangers don't like the extra weight.

Ralph Doncaster.

structurally integrated insulation

Some building materials incorporate foam board insulation into structural components for above and below grade uses. Structurally insulated panels (SIPs) are made of foam board insulation sandwiched between two outer layers of structural sheathing material. The resulting panels can be used as a structural system or as infill in a post and beam structure. These units are made in different sizes according to the requirements of the project.

SIPs are factory-built units and can be manufactured to strict standards. Case studies have shown that properly installed, SIPs can reduce air infiltration to minimal levels. Because SIPs homes can be built faster than conventional stick-built homes, the costs of carrying a construction loan can be reduced.

Insulating Concrete Forms (ICFs), can be used above and below grade, and come in three basic configurations: hollow foam blocks, foam planks held together with plastic ties, and panels held together with plastic ties. The foam component is typically expanded polystyrene (EPS) or extruded polystyrene (XPS). Insulation values vary with the material and its thickness. ICF foundation walls

Photo courtesy of Rob Dumont, Saskatchewan Research Council

are faster and easier to construct than concrete masonry units or poured concrete walls. The insulation materials, when finished with manufacturer's specified drainage plane material, are a near guarantee that there will be minimal moisture concerns.

Another type of ICF uses wood-fibre composite forms to provide both thermal and sound insulation. The forms are available with mineral wool insulation to provide an energy-efficient, fire and termite resistant wall system. This system is capable of absorbing high levels of airborne moisture without damage, and does not support fungal or mold growth.The material does not give off any toxic runoff or airborne pollutants and does not require a fire-rated finish, as the material itself is fireproof.

assemblies

One of the best ways to get a super-insulated wall on your new house is a 'stand-off' wall, where you have an exterior load-bearing wall, standard stud construction, and you build another, non-load bearing stud wall several inches away (that's the 'stand-off') from the first wall. This provides a deep cavity that can be filled with inexpensive fibrous insulation. You lose a little floor area, or accommodate the extra wall by using a bigger footprint for the building. If you use batt insulation you can install three layers once the studs are in place. Place the first one vertically in the exterior wall cavity, the second is stacked horizontally between the two stud walls and the third one, like the first, goes into the wall cavity vertically.

In an existing house, it can be very difficult to build a new wall inside the old one, but you can certainly build a new wall on the outside. This exterior stand-off wall is commonly called a 'Larsen truss'. A lightweight, non-load bearing truss is built on site and attached to the existing wall. The cavity can be filled as for an interior stand-off wall.

Layers of rigid board insulation can be added to the exterior or interior of walls to increase insulation value as well.

LARSEN TRUSS

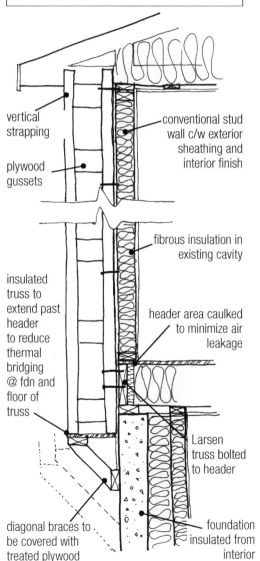

vertical strapping

plywood gussets

insulated truss to extend past header to reduce thermal bridging @ fdn and floor of truss

diagonal braces to be covered with treated plywood

conventional stud wall c/w exterior sheathing and interior finish

fibrous insulation in existing cavity

header area caulked to minimize air leakage

Larsen truss bolted to header

foundation insulated from interior

INTERIOR FOAM
gypsum board sealed as air barrier
Type IV or polyiso rigid foam board insulation
strapping @ stud spacing
Type IV or polyiso rigid foam board insulation
framed wall w/fibrous insulation in cavity
sheathing
housewrap (if required by code)
vertical strapping for rainscreen
solid wood siding

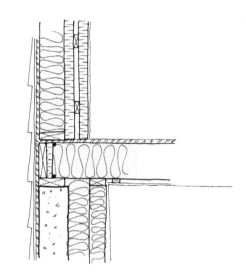

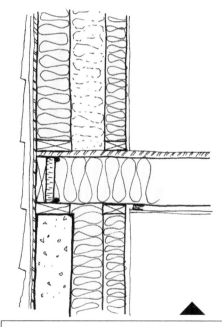

INTERIOR STANDOFF
gypsum board
stand-off wall c/w insulation batt.
VDR, taped and sealed to back of stud wall insulation
cavity c/w batt installed horizontally
exterior bearing wall c/w batt insulation installed vertically
plywood sheathing
housewrap (if required by code)
vertical strapping for rainscreen
horizontal siding

EXTERIOR FOAM
gypsum board sealed as air barrier
w/fibrous insulation in cavity
plywood sheathing
Type IV or polyiso rigid foam board insulation
strapping @ stud spacing
Type IV or polyiso rigid foam board insulation
housewrap (if required by code)
vertical strapping for rainscreen
horizontal wood siding

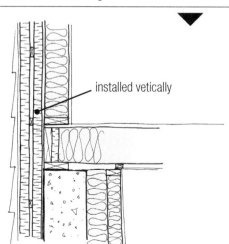

installed vetically

In any situation where you are adding insulation, remember that the VDR must be at least two-thirds in from the cold side of the wall to prevent condensation inside the wall cavity.

If you use dense-pack insulation, install cardboard baffles between the inside and outside studs to create a cavity into which the insulation can be blown.

121

windows and skylights

WINDOWS

Nearly half of our physical comfort is based on the radiant heat exchange between our bodies and the surrounding surfaces in a home. As up to 25 per cent of the exterior wall surface can be devoted to the weakest player in the thermal envelope – glazing – high performance windows are key to overall comfort levels in a home. Over the last 10 or 15 years, windows have improved and costs have gone down. Double-paned windows with low-e coatings and gas fill have become commonplace for new construction and the replacement market. Newer, better performing windows are now on the market as materials and technologies improve.

In general, windows with lower U-factors (higher R-values) perform better in heating-dominated climates, and windows with lower solar heat gain coefficients (SHGC) perform better in cooling climates. Air tightness is another important factor.

'High Insulation Technology' (HIT) windows improve the design and performance of today's energy-efficient windows. Some use low-e films suspended between two panes of glass to create two or more 'interpane' air spaces. These windows achieve better performance than triple pane windows and have lower weight. Lower weight means lower cost for shipping and the elimination of weight considerations related to installation or replacement of hefty triple pane units.

Other HIT strategies include vacuum windows and aerogels between panes. The best windows currently on the market use a combination of suspended films, low-e coatings, low-iron glass, gas fill, pultruded fibreglass frames insulated with vacuum-silica aerogel, and low-conductivity spacers.

WINDOW ELEMENTS DEFINED

U-FACTOR

The U-factor is the measure of the rate of non-solar heat-flow through a material or assembly. It is expressed in units of BTU/ or W/sq. m x °C (hr. x sq. ft. x °F), and may be expressed for the glass alone or for the entire window assembly, including frame and spacer materials, so make sure you are comparing apples to apples when looking at U-factors. The lower the U-factor, the greater a window's resistance to heat-flow and the better its insulating value.

GAS-FILL

An inert gas replaces the air inside a glass unit to reduce temperature transfer and slow convection. Argon gas, which is most commonly used, is inexpensive, nontoxic, non-reactive, clear and odourless. Krypton gas, which is about 12 times more dense than air is also used, but it is more expensive. A mixture of krypton and argon gases is also used as a compromise between thermal performance and cost.

When choosing windows, be sure to consider the relationship between the glazing area and the overall window area. Frames and mullions can account for up to 40 per cent of the total window size!

Several factors define a high-performance window: framing materials that don't conduct heat, multiple panes of glass, low-emissivity and solar control coatings on at least one surface of the glass, low-conductance gas fills between panes, thermal breaks at the edge spacers and edge sealing. These factors vary depending on your climate – some of them even vary depending on the orientation of the window itself.

CHARACTERISTICS OF DIFFERENT GLAZINGS

*R-value is for centre of glass
U-vaue measured in W/m²K
‡the higher the solar heat gain coefficient (SHGC), the greater the percentage of solar energy admitted

type of glass	R-value*	U-value	solar heat gain‡
double pane	2.0	2.84	.78
double, low-e	3.0	1.88	.75
double, low-e, gas-filled	4.0	1.43	.75
triple pane	3.1	1.85	.70
triple, low-e	3.9	1.47	.66
triple, low-e, gas-filled	6.7	0.85	.66

SOLAR HEAT GAIN COEFFICIENT (SHGC)

SHGC is the fraction of solar radiation transmitted through a window or skylight, expressed as a percentage. The lower a window's SHGC, the less solar heat it transmits and the greater its shading ability. SHGC can be expressed in terms of the glass alone or can refer to the entire window assembly.

SPACER

A spacer is a material placed between two or more panes of glass in an insulated glass unit to bond and seal the glazing unit. High-performance windows use 'insulating' spacer materials (e.g., rubber vs. stainless steel or plastic vs. aluminium) and/or a thermal-break spacer design to reduce conductivity between interior and exterior glass panes.

UP AND COMING WINDOW TECHNOLOGIES

As part of a 'smart' integrated system, active window insulation (automated venetian blinds) can be used as a daylighting strategy to reduce lighting and cooling loads by 14 to 28 per cent. The major disadvantage of integrated blinds is the obstruction of views during daylight hours. To get around this, some manufacturers are introducing eletrochromic glazings ('active glazings') that achieve similar reductions in lighting and cooling loads while maintaining views. A thin film on the glass uses a small DC current to darken the window for shading or lighten it for solar gain. Some units switch from one mode to the other, while others have intermediate switching. Eletrochromic (EC) units can be tied into wireless control systems and can be powered by PV panels. Both of these technologies have been been introduced as a commercial energy efficiency measure, but as the technology is refined, EC windows will come into homes.

LOW-E GLASS

The 'e' refers to emissivity, which is the relative ability of a surface to reflect or emit heat by radiation. Glass has low emissivity due to a film or metallic coating on the glass or suspended between two panes of glass to restrict the passage of radiant heat. The lower the emissivity, the less heat that is emitted through a window system. Emissivity is typically measured by U-factor (which is the inverse of a material's R-factor). There are generally two types of low-e coatings for glass.

Hard coat, or pyrolytic coating, is sprayed onto the glass surface during the float glass process and is considered to be a medium performer. One of the earliest coating types, this type of low-e coating allows a high level of solar heat gain and is tough enough to be exposed to the elements. It is most appropriate where solar heat gain is desired.

Soft coat, or sputter coating, is applied to the glass in multiple layers of optically transparent silver sandwiched between layers of metal oxide in a vacuum chamber. This process provides a higher level of performance and a nearly invisible coating. Spectrally selective sputter coat offers the best performance in terms of reducing solar heat gain. Soft coat low-e is more fragile than hard coat, and has to be applied to one of the inner surfaces of the glass panes.

In heating dominated climates, the low-e coating is placed on the interior glass surface (surface 3) of a double-pane window. This allows solar radiation to pass through the exterior pane, contributing to heat gain during the winter, but interior heat is reflected back towards the living area. In cooling dominated climates, the low-e surface is placed on the exterior pane (surface 2) to reflect solar radiation away from the window. Triple pane units can have one or two low-e coatings on any of the interior glass surfaces.

To make best use of differing levels of sun exposure, you can specify different coatings for windows having various locations in the same house. This variety of coatings can increase efficiency, but windows with different coatings may appear to have slightly different coloured glass.

For updates on windows and other new products, check out the Building Green product directory and their annual lists of top products at www.buildinggreen.com

OVERALL ENERGY REDUCTIONS

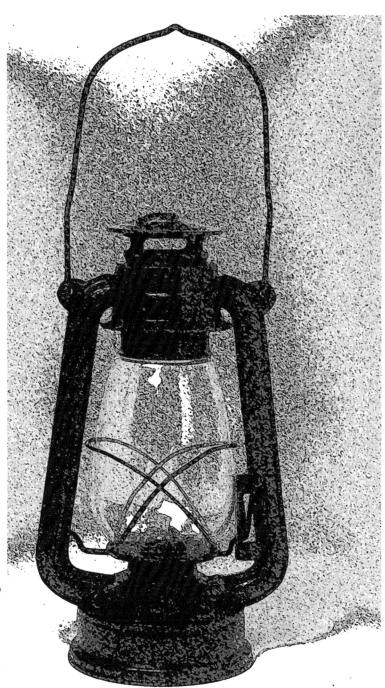

reducing overall energy use

Your house uses energy in three major categories: space heating, water heating, and lighting and appliances. You can minimize your space heating requirements in many ways through planning and design. Energy needed for water heating and for lighting and appliances can also be reduced through planning and design, but how you use lighting and appliances has as many variables as there are people in houses: these loads are ultimately dictated by the efficiency of the devices themselves, along with your lifestyle.

The electrical loads from appliances, lighting and miscellaneous electric loads (electronics, fans and gadgets) is often called 'secondary energy use'. In combination, these are also referred to as the house 'baseload'. In general, the loads associated with secondary energy uses are highly variable, being based on occupant lifestyle and product choices, but there are ways to significantly reduce them. The first thing you need to do is start thinking of how you would live if your home was off-grid.

If you make your own electricity, you are on a strict energy budget. Just like the dollars in your income, there are only so many kilowatt hours to go around.

thinking off-grid

If you were to go off-grid – to supply all of your electricity – how big a system would you require to carry on your current lifestyle? With electronics.

Many off-grid houses in northern climates operate comfortably with all the 'modern conveniences' using on-site generation systems (wind, PV, or a combination of both, with a backup generator) ranging in size from 2kW to 10kW capacity, producing between 0.5 kilowatt-hours and 5 kilowatt-hours a day (based on locations that produce at least 100 kilowatt hours annually for each 100W of PV installed). In terms of electricity, many grid-connected houses are on the 'supersize me' diet! Even if you are not planning on installing your own PV or other energy-generating system, you can still benefit greatly from cutting your baseload.

So how do you drop this excess? The most important thing to recognize is that your baseload is the major area in which your day-to-day choices are the leading factor. Put simply, you need to choose to use less energy. Here are some ways you can do this.

appliances

Ironically, as houses have become more energy efficient, the average size of a new home has increased. The number of electronic devices has multiplied, and the average household is using *more* energy than ten years ago.

Major appliances, like furnaces and water heaters, are commonly replaced when they fail. When this happens, all too often the decision about what to replace them with becomes a no-brainer: whatever is immediately available. Most of us don't have the opportunity to make a decision based on research and comparison shopping at this point. Perhaps there are a few features that you want, a colour, a style, but most likely, the governing factor is the purchase price. So while you can't really predict exactly when your appliance is going to die, if it's put in seven to ten years of service, it's probably due for replacement. At this age most modern appliances are closing in on the end of their useful life. Because appliances manufactured 7 to 10 years ago use more energy than just about any new appliance in today's market, you should be considering replacing them anyway. Therefore, doing a little research prior to an appliance emergency will save you money in the long run.

To make the most of your new appliance purchase, topping your list of wanted features should be an energy rating. In the US, look for the Energy Star™ label. This means that the appliance is in the top 30 per cent of the category when it comes to energy efficiency. And just because an appliance is Energy Star™ rated doesn't make it energy efficient when compared to smaller models or models with fewer whizbang features.

In the UK, appliances are graded from A to G, an A-rated machine provides the biggest CO_2 saving, and can reduce the long-term running costs too. The most efficient models get an Energy Saving Recommended (ESR) badge. Fridges and freezers go up to an A++ rating. Be wary when looking at fridge and freezer ratings – they are based on volume. This means huge refrigerators can get an A++ rating, even though they use nearly 50 per cent more energy than a typical 300-litre model.

While most Energy Star™ or ESR models will have a slightly higher purchase price, it is safe to say that the higher price will be paid off within a year or two. Most appliances have a ± 10 year lifespan, so you have several years to benefit from the cost savings. Energy efficient freezers, fridges and combination units, dishwashers, washing machines, fans, space heating and cooling equipment are all readily available.

FREEZER: Chest freezers are more energy efficient, although they do take up more floor space. Upright freezers allow the cold air to spill out of the whole freezer compartment for the entire time the door is open, whereas a chest freezer has little cold air spillage. Tip: keep your freezer as full as you can. If the power goes out, stuff the empty space within the freezer with pillows and blankets to act as insulation. During a power outage, a chest freezer can be 'charged' by a generator for 1 hour every 24 to 48 hours if you don't open the lid! This will keep the food solidly frozen. Chest freezers can be put on timers so that they don't cycle on and off as much as the manufacturers' settings require. This is an especially good point for those on time-of-use programmes, or who otherwise have to manage their energy use.

REFRIGERATOR: If you don't have a well-pump, your refrigerator is probably the biggest contributor to your electricity use. Buying the most efficient refrigerator will mean that you're emitting the least amount of greenhouse gases and CO_2 into the atmosphere, as well as saving money. For the same reasons as noted above, fridges with top or bottom mount freezers are more efficient than side-by-side units. Models with bottom-mounted freezers are more efficient than those with top-mounted freezers. Bottom-mounted, solid-sided freezers on drawer slides are more efficient than those with swing-out doors and pull-out baskets and shelves. Features such as icemakers and water dispensers are conveniences that add to the purchase price and to energy use. A pitcher of cold water in the fridge compartment is less costly. Side-by-side models can use up to 13 per cent more energy than similar top-freezer models, and automatic ice-makers increase energy use by up to 20 per cent.

DISHWASHERS: There has been quite a bit of debate over energy efficiency and water conservation when comparing machine- vs. hand-washing. This was partly because older model dishwashers require high domestic hot water temperatures and because they use a lot of water: 30 to 57 litres (8–15 gal) per normal cycle in 10-year old units. New energy efficient machines typically use 15 litres (4 gal) a load, and 40 per cent less energy than non-qualified models. Most models have a little heater coil in them to boost the water temperature, allowing the domestic hot water settings (for the whole house) to be lowered. Features to look for include:

- Several cycle selections, including a light or energy-saving cycle and water conservation cycles.
- An air-dry feature, which cuts down on energy use.
- Make sure you pick a model that fits your household: a compact capacity model holds eight place settings and six serving pieces, while standard capacity models hold more. And of course, don't run it until it is full!

WASHING MACHINES: Generally, washing machines are second only to toilet flushing for residential water use, using up to 150 litres 40(gal) a load! Energy efficient appliances use between 68 and 95 litres (18 and 25gal). Front-loading machines are, in general, more efficient than their top-loading counterparts, using 30 to 50 per cent less energy. They also have the benefit of faster spin times, which means clothes dry faster. Due to their increased efficiency, some local utilities offer rebates for purchasing front loaders. Wash your clothes in cold water. 90 per cent of the energy consumed by a washing machine is used to heat water. In situations where hot water is necessary (for instance, to kill dust mites in bedding), use cold water to rinse. Use high-efficiency liquid detergent in top loaders. Features to look for include:

- Lower 'water factors' (the amount of water used in relation to the size of the drum)
- Variable spin cycle
- Self-adjusting water levels
- Delayed start/programmable start (good for time-of-use programmes or off-grid/grid-connected systems where load management is a factor)

Although clothes dryers are much more efficient than they used to be, a solar powered exterior textile moisture removal device – a.k.a. a clothesline – has the best energy rating around!

lighting

Lighting is a relatively small load in the house, but does impact on your energy use. One of the best ways to reduce your lighting energy use is to take advantage of natural sunlight and bring it into every room of the house. Different spaces in a house have different lighting requirements, but often the lighting in those spaces is in the wrong place for the tasks and activities that occur in each area. Once you have determined how much light you need and where you need it, installing energy efficient bulbs and fixtures will drop lighting energy use further. In addition, occupancy sensors can save you up to 50 per cent of the energy used by each light bulb controlled by the sensors. This also prolongs the life of the bulb, saving more money.

The goal is for daylighting to satisfy your general, or ambient, lighting needs. As a result, only task lighting will be needed during the daytime. Minimal electric ambient lighting will be needed only at night or on very dark days. Windows, skylights and tubular skylights as well as interior windows and openings all help increase daylight. Painting interior walls and ceilings in light colours can help distribute daylight into the house.

See the section on house design for more details and strategies for integrating daylight.

You can replace a regular 60W incandescent bulb with a 52W energy-saving incandescent, a 15W compact fluorescent, or replace the existing fixture with one that holds two 10W Warm White fluorescent tubes. All of these offer the same level of light (lumens).

LIGHT LEVELS
Determining exactly how much light you need in a space can be bewildering. When you purchase a lightbulb, the package will typically show the wattage of the bulb, the rated hours of use and the level of light it will provide, typically in lumens. A lumen is a measure of lighting efficiency that relates the amount of light (illuminance) to the watts you pay for. The more lumens, the brighter the light. Following are two tables showing some lighting levels for individual rooms and for tasks. This will help you determine how much light you need, and where you need it. First let's look at how much light the sun can give us.

A quick look at the following table shows that the sun can provide between 75 and 100 per cent of the ambient lighting needs of most spaces during the day, regardless of a clear or cloudy sky. With a little help from design strategies that help 'bounce' light through rooms (without producing glare), the sun can easily illuminate most homes during the day. If an existing home does not offer very much leeway in terms of daylighting, tubular skylights work very well to improve ambient lighting without requiring a large area of glazing on the roof.

Multi-use spaces, such as living rooms, need adaptable lighting. Recessed can lights (also called pot lights) provide directional light. Wall sconces and cove lighting provide ambient light. Task lighting is usually provided by portable

LIGHT LEVELS FROM THE SUN.

Point of reference	Value
Sunlight, outside of glazing	80-110 lumens
Overcast or clear sky, outside of glazing	105-140 lumens
Sunlight, inside of glazing	75-225 lumens
Overcast or clear sky, inside of glazing	100-290 lumens

lamps. However, be warned! Where recessed lights are installed in an insulated ceiling, they need to be approved for 'insulation contact'. They must be properly sealed to prevent air leakage and moisture migration into your attic space. It is better to avoid installing recessed lights into the building envelope.

Task lighting strategies put a brighter light source close to the task, allowing general, or ambient, light levels to be lower. Mix ambient and task lighting according to the space. For ambient lighting in general, use low-mercury fluorescents mounted in indirect fixtures (sconce, recessed, surface-mount, indirect cover, etc.). Nearly all styles of fixtures are available for straight-tube or compact fluorescent lamps. In the kitchen and bathroom, use either recessed compact fluorescent lamps (CFLs) or LED light fixtures for task lighting.

Room Lighting Levels	
Kitchen	
General	300 lumens
Counter top	750 lumens
Bedroom (Adult)	
General	300 lumens
Task	500 lumens
Bedroom (Child)	
General	500 lumens
Task	800 lumens
Bathroom	
General	300 lumens
Living Room/Den	
General	300 lumens
Task	500 lumens
Family Room/Home Theater	
General	300 lumens
Task	500 lumens
TV viewing	150 lumens
Laundry/Utility	
General	200 lumens
Dining Room	
General	200 lumens
Hall Landing	
General	500 lumens
Home Office	
General	500 lumens
Task	800 lumens
Workshop	
General	800 lumens
Task	1100 lumens

The basic rules of thumb for planning your lighting design are as follows:
#1: Use interior windows and openings as well as tubular skylights to bring daylight into all rooms
#2: Use high-efficiency task lighting for work areas
#3: Use indirect fixtures for general 'ambient' lighting and 'bounce' the light off the ceiling
#4: Install occupancy sensors in hallways, bathrooms and other areas that are used for relatively short periods of time.

LIGHTING THROUGH THE AGES

Along with joints and other body parts, our eyes deteriorate as we age, and so a part of our at-home safety is directly linked to lighting. As we age, our lenses get thicker, and more light is needed to get to the back of the retina. There is also a change in our perception of colour: we see more amber, which absorbs bluish and purplish colours. So, older adults need light sources that give off cooler, bluer light – such as fluorescents and blue LEDs – rather than incandescent bulbs, which tend to have a more yellow cast. And as our lenses get thicker, we also become more sensitive to glare because the additional layers scatter the light we see.

A good way to increase overall light levels is to install linear fluorescent tubes near the ceiling. The light, bounced up on light-coloured walls, increases the amount of indirect light in a room. Under-cabinet lighting can also increase overall light levels. However, you should only use under-cabinet lighting if the countertop surface is light-coloured and has a matt surface, or else the light may cause too much glare.

Task lighting, in tandem with overhead lighting, is especially important for older adults, who have a harder time adjusting their eyes to varying light levels. To improve task lighting, install a recessed light above the kitchen sink. In the bathroom, a wet-location-rated downlight can provide good task lighting in the shower. At the vanity and mirror area, covered fluorescent tubes on either side of the mirror light both sides of the face. If the light is installed just above the mirror, it creates a shadow, interfering with personal grooming such as shaving and applying make-up.

Safety tip: since older adults get up frequently in the middle of the night to use the bathroom, plug in an amber-coloured LED nightlight.

Lighting Levels by Task	
Casual reading	300 lumens
Writing	500 lumens
Studying	1000 lumens
Food preparation	750 lumens
Cooking	300 lumens
Dining	200 lumens
TV viewing	150 lumens
Laundry	200 lumens
Detailed work	1000 lumens

phantom loads

Phantom loads are generated by 'always on' remote control devices, any piece of electronic equipment that has a sleep mode, and any appliance that has a digital clock in it. Any device with an LED power indicator, which obviously uses power itself, is generating a phantom load. Also called vampire loads and stand-by energy losses, they have increased considerably in recent years. For any single appliance in your house, the phantom load is unlikely to be very large – at worst, 15 to 20 Watts – but as most homes have many of them, they really add up! In fact, phantom loads can account for up to 11 per cent of your annual electrical power consumption, and can account for up to 75 per cent of the energy consumed by the appliance in a year. According to the January 2008 Wikipedia entry for phantom loads, some studies have suggested that the total phantom load caused by the United States alone would provide enough power to handle the electrical needs of Vietnam, Peru and Greece!

An effective low cost strategy for dealing with this problem is to plug appliances into a common power bar. When the power bar is turned off, the appliances no longer draw power from an outlet, thereby eliminating phantom loads. The challenge here is that most power bars are unsightly and get stuffed behind furniture, making them awkward to turn on and off. This renders them pretty well useless.

Another option, for about the same price, is a remote control outlet system, which comes with two or more outlet receivers. The receiver goes in the wall outlet and a regular power strip plugs into that reciever, allowing remote control over all the devices on that strip. A little battery powers the system, but compared with the savings it is potentially a major gain. You just have to remember to hit the off button when not using the appliances. You can chase down your phantom loads with a 'Kill-a-watt' meter. This is a fun way to get children involved in reducing energy needs.

'Smart meter' for grid-connect system.
Photo courtesy of Rob Dumont, Saskatchewan Research Council.

Home energy monitors can also be used to point out the energy use (and financial costs) in your home. Pilot projects and demonstrations have shown that energy monitors can actually modify energy use patterns, with 10 to 25 per cent reductions in energy loads. That's pretty significant! Off-grid homes and homes with grid-connected PV systems have built-in energy monitors in their systems. There have been some barriers to bringing home energy monitoring systems to market, but improvements to wireless technology and the emergence of wireless standards may soon make customized energy monitoring and automation solutions standard in both new and existing homes. There are, however, some unresolved issues with wireless systems and exposure to electromagnetic fields (EMFs) in the house.

Some public libraries have partnered with governmental energy departments and ministries to provide kill-a-watt meters on a loan basis – check it out!

reducing hot water use

Heating water can account for 15 to 20 per cent of the energy use in a home. Water heating is second only to space heating in terms of energy use in most cool climate homes, so it makes sense to keep track of your hot water consumption and make sure that the heater is running as efficiently as possible.

Standing losses from electric water tanks can be reduced significantly through measures such as insulation blankets. Many hot water tanks are pre-insulated 'conserver' models, so adding a blanket may not make a difference.

When designing a new home or moving a water heater, locate the unit as close as possible to the kitchen, laundry and bathrooms to reduce the heat loss to long pipe runs.
Natural Resources Canada reports that reducing the length of a hot-water pipe from 10 to 3 metres (33 to 10ft) will save enough energy in one month to heat water for 10 showers. Similarly, small diameter pipes are more energy efficient than bigger pipes, which carry larger amounts of hot water and so lose more heat.

Insulate the first 3 metres (9ft) on cold water pipes and the first 2 metres (6ft) on hot water pipes running to and from your water heater. Do not place any pipe-wrap insulation within 15cm (38in) of exhaust vents at the top of water heaters, and never insulate plastic pipes.

Many water heating tank manufacturers pre-set the temperature of the tank to 60°C (140°F). You can lower the thermostat to as low as 55°C (130°F) to save energy. Do not set it any lower, as this risks the growth of bacteria that are hazardous to health, such as legionella.

ENERGY AND WATER CONSERVATION
If you use water provided by your municipality, the water is usually pumped from a source and treated with chemicals before you use it. Then it is treated again before it is put back into the environment. All this movement and treatment of water takes energy, and producing this energy contributes to greenhouse gas emissions. The use of electricity or natural gas for your water heater further adds to greenhouse gas emissions.

As a house becomes more energy efficient, standing losses from water heaters situated on the main floor of a house or in a finished basement area can be considered a viable part of the heating regime in the house. No longer being lost to an uninhabited basement, the standing loss becomes a useable internal gain.

DRAINWATER HEAT RECOVERY (DWHR)

This is an elegantly simple technology that recycles heat from wastewater. DWHR units take advantage of the fact that water clings to the inside surface of a vertical drainpipe due to surface tension. Typically, DWHR units are made from a length of copper drainpipe wrapped with small diameter copper tubes. As cold water circulates through the small tubes, it is heated by the drainwater clinging to the inside of the drainpipe. These types of units work best with simultaneous flows from showering, effectively reducing your showering hot water requirements by 25 to 40 per cent. As showering can account for nearly a quarter of your overall hot water use, this can result in a significant reduction. Make sure the DWHR unit is installed on the drainpipe that services the most-used shower in the house.

Drainwater heat recovery units.
Photo courtesy of Natural Resources Canada, Buildings Group.

AT THE TAP

The first steps looked at ways to improve the efficiency of the water heater. The next step is to look at ways to reduce your hot water use. Number one on this list is to fix leaky fixtures (a one-drop-per-second leak equals 16 baths per month or 160 cycles on an automatic dishwasher!).

Energy-efficient low-flow showerheads conserve energy by up to 60 per cent without changing water pressure. A flow rate of two litres per minute will reduce water consumption while keeping the shower enjoyable. A shut-off button on a low-flow showerhead allows you to be very water efficient – you can interrupt the flow while you lather up or shampoo and then resume at the same flow rate and temperature. By taking five minute showers instead of baths; you'll use up to 50 per cent less hot water. If you run the tap while shaving or washing your face, money and energy is going down the drain. Partially fill the basin with hot water – you'll save a lot of hot water.

Rinsing dishes under the tap also wastes a lot of water. Rinse your dishes in a large bowl of water, partially fill one side of a double sink, or slowly pour a bowl of water over dishes after putting them in the drainer. energy efficient dishwashers use less energy and water than washing dishes by hand.

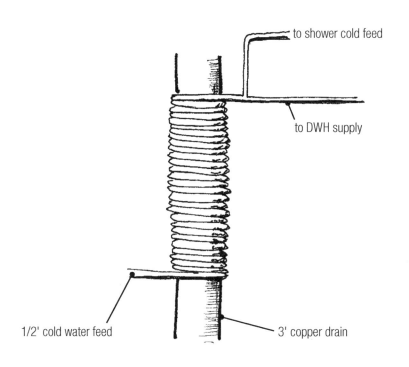

Schematic of most efficient plumbing for a drainwater heat recovery unit.

to shower cold feed

to DWH supply

1/2' cold water feed

3' copper drain

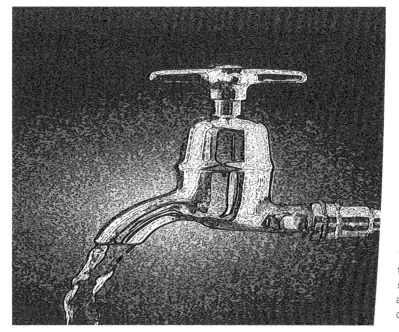

Where the risk is greatest to children, the infirm or elderly, tempering valves are available for individual taps. A professional installer will be able to provide you with specific details about the best way to do this in your home.

Turn down your water-heater thermostat to its minimum setting when you plan to be away for extended periods of time.

SOLAR ADD-ONS

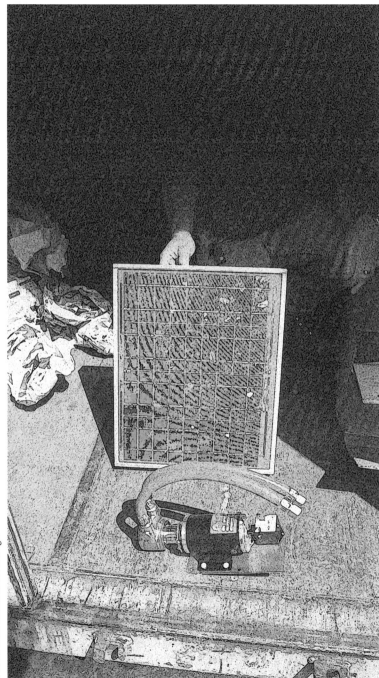

Solar add-ons can be incorporated into nearly any home during construction or, with some planning, after construction. These systems are defined as 'active' indirect solar systems, because they rely on mechanical means to transfer heat (pumps or fans) or store energy (inverters and batteries). Typically, there are two main categories: solar thermal (to heat water or air) and photovoltaic (to make electricity). This section will introduce several different types of systems, how they work and what they can contribute to your energy requirements.

Liquid-based solar thermal systems can provide heat for water and/or space heating systems. A standard solar domestic hot water (SDHW) system can provide 50 to 70 per cent of a typical household's needs. Adding more storage and reducing your hot water usage can improve the solar contribution.

Air-based solar thermal systems provide pre-heated ventilation air or forced-air space heating. Solar air heaters and pre-heaters have a proven track record in commercial and industrial settings, but they are relatively new to the residential market. The installation, operation and performance of these systems varies with the manufacturer.

Photovoltaic (PV) systems provide solar-generated electricity to remote, 'off-grid' homes, or supplement utility-supplied electricity in 'grid-connected' homes.

The house design and planning section discusses ways to make your home 'solar ready' for both thermal and PC applications.

solar thermal systems

SOLAR HOT WATER

There are three basic types of solar hot water systems in use today: closed loop, open loop and thermosyphon. In most cool climates, where there are freezing conditions, the closed loop system is common. Another type of system is a 'spatial' solar hot water system, which can be a viable option if your house has a tall greenhouse space.

Solar domestic hot water (SDHW) systems, offered by many distributors in most markets, are designed to heat water for cleaning, showering and bathing. These systems typically have 2 flat-plate collectors or 30 evacuated tubes. They are installed in tandem with a conventional water heater (electric, gas or oil). The conventional water heater comes on when the water from the solar storage tank is not up to the required temperature. The water going into the conventional tank is pre-heated to a certain extent every day, depending on the amount of solar gain. On clear days in the summer, the SDHW system can contribute upwards of 90 per cent of your hot water needs. When the sky is overcast, or when the temperature is below 0°C, the solar contribution drops. Annually, a SDHW system can supply 50 to 70 per cent of your hot water.

COLLECTOR TYPES

Collectors, or absorbers, transform sunlight into heat. There are two main types of collectors that are useful for residential applications: the glazed flat-plate and the evacuated tube. Making a decision on which type of collector to install is challenging, as distributors of each type make competing claims about efficiencies. See page 138 for more information.

Typically, you will get the best return for your investment in a SDHW system if you use a large amount of hot water. A family of four will have a shorter payback period than a single person, simply because of the volume of water they require.

Showering or washing in the afternoon or early evening uses water heated by the sun. Many clothes and dish washing machines have a delay start function that can also be used to time hot water use with maximum solar gain. If you are on a time-of-use program you benefit both ways, as mid-afternoon is often a 'shoulder' time with a higher per-kilowatt price than off-peak.

Photo courtesy of Natural Resources Canada, Building Group.

INTERNATIONAL STANDARDS RELATING TO SOLAR WATER HEATING

ASHRAE 90003 Active Solar Heating Design Manual
ASHRAE 90336 Guidance for Preparing Active Solar Heating Systems Operation and Maintenance Manuals
ASHRAE 90342 Active Solar Heating Systems Installation Manual
ASHRAE 93 Methods of Testing to Determine the Thermal Performance of Solar Collectors
Solar Rating and Certification Corporation (SRCC) OG-300-91: Operating Guidelines and Minimum Standards for Certifying Solar Water Heating System
Relevant European standards include EN 12975 (for solar thermal collectors) and EN 12976 (for factory made solar thermal systems)

FLAT-PLATE COLLECTOR (FPC) CUTAWAY

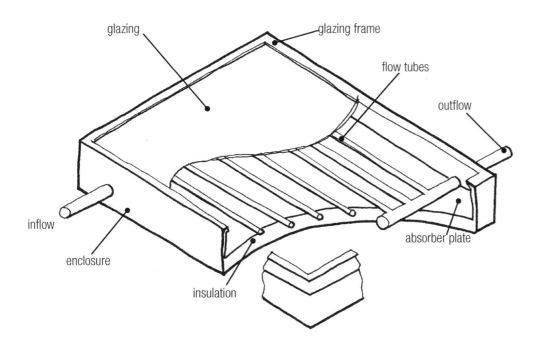

glazing

glazing frame

flow tubes

outflow

inflow

enclosure

insulation

absorber plate

FLAT-PLATE LIQUID-BASED COLLECTORS

A glazed flat-plate collector is a shallow rectangular box with a flat black plate behind a tempered glass cover. The plate is attached to a series of parallel tubes or one serpentine tube through which water or antifreeze solution passes. Fins of anodized copper or other heat collecting material are used to improve performance. As the liquid circulates through the system, it absorbs the heat from sunlight.

HEAT-PIPE EVACUATED TUBE SOLAR COLLECTOR DIAGRAM

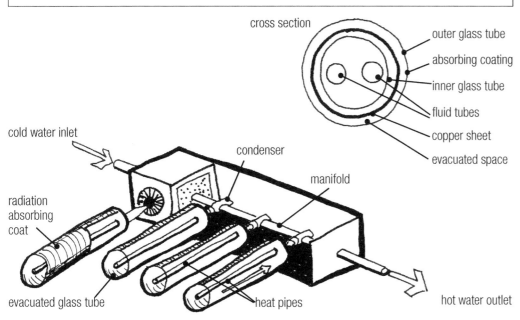

cross section

outer glass tube

absorbing coating

inner glass tube

fluid tubes

copper sheet

evacuated space

cold water inlet

condenser

manifold

radiation absorbing coat

evacuated glass tube

heat pipes

hot water outlet

EVACUATED TUBE COLLECTORS

An evacuated tube collector is made up of several individual glass tubes, each with its own heat pipe suspended in a vacuum (that's the 'evacuated' part). The pipe transfers the heat absorbed from the sunlight to a condenser through the top of each tube. The condensers are attached to an insulated manifold. The manifold collects the heat from the condensers and transfers it to a heat exchanger.

For insurance purposes, installing equipment that meets standards is the best option for most home owners. Lending institutions may also look more favourably on a system that meets a recognized standard.

NOTE
Your choice of FPC or ETC
will be affected by how and
where you plan to install the
system.

FLAT PLATE VS. EVACUATED TUBE

There is an ongoing debate about which collector is more efficient. The fact of the matter is this: under specific circumstances, flat plate collectors outperform evacuated tube collectors and vice versa. Neither is 'more efficient' than the other.

The biggest impacts on the performance of any thermal collector are:

1) The difference between the collector inlet temperature and the outside ambient air temperature

2) The amount of available radiation

3) The match between delivered temperature and the required end-use

Flat Plate Collector	Evacuated Tube Collector
Best suited to applications where delivered temperature is lower than 93°C (200°F)	Best Suited to applications where delivered temperature is higher than 93°C (200°F)
Currently, FPC have a better relation between supplying DHW efficiently and the cost to install, however, this could change as ETC prices go down	ETC are more expensive than FPC, but the price point is changing (nearly all of the 27 million homes in China that have solar hot water use inexpensive ETC systems)
Perform best at 14 to 50°C (25 to 125°F) above ambient air temperatures	Perform best at 50°C+ (125°F+) above ambient air temperatures
Best suited to milder climates and low-temperature applications such as pool heating, domestic water heating and radiant floor space heating	Better suited to cold and cloudy climates and higher-temperature application such as space heating using baseboard radiators and industrial process hot water or steam
Built within a solid, sealed case and covered with tempered glass for protection from the elements; the air space in the case leads to conduction and heat losses on cold and windy days	The glass tube is hermetically sealed within a vacuum that eliminates convection and conduction heat losses and also isolates the collectors from weather conditions: vacuum can be difficult to maintain in some types, especially those with metal to glass seals
Built to be drained completely, allowing unit to be used in open loop or closed loop systems	Most trap significant amount of fluid in the manifold that cannot be drained by gravity, only suitable for closed loop systems. Servicing fluid in system can be more complicated
No internal overheating protection methods, rely solely on external limiters. If these fail, the collectors can be damaged	The heat pipe is safeguarded from overheating by the conduction properties of the transfer fluid in the heat pipes
Must be installed as one solid unit. If a portion of the collector fails, the whole panel must be shut down and retrofitted or replaced	Can be carried in pieces for lighter, easier installation. If one tube is damaged, only that tube needs to be replaced (but noticing that one tube is damaged requires inspection)
More dependent on orientation and angle of installation	More flexibility in placement because of the concave surfaces of the tubes
Concerns about snow cover in low-slope applications	Concerns about snow build-up between tubes
Many Canadian and US manufacturers	No North American manufacturers (at time of printing), made in Europe, Australia and China

If you are thinking of using solar thermal for space heating, look into the economics of an evacuated tube system, installed vertically on a wall with a suitable overhang for summer shading.

138

In many western countries, there are significant tax breaks and incentives for installing green types of energy.

When considering the cost of installing a residential solar energy system, you should check out and include in your calculations any tax you can reclaim or subsidies which are applied by local or national government.

In order to work out how much of your hot water can be generated in your specific house, taking into account where the panels could be located, you will need to get a quote of the estimated energy which could be generated from a particular system on your house per year.

Obviously these estimates can't be exact, as it depends on the sunshine. Solar energy providers tend to give a range of plus or minus 20 per cent. A typical solar domestic hot water system will generate about 1,500kWh-e of useful heat annually (resulting in a reduction in CO_2 of between 325 and 700kg per year, depending on your current fuel type).

When sizing a system, estimate that you will need 1m² to 2m² (10 to 20ft²) of collector per person to meet about half of your hot water demand. Your system will need some storage, optimized to the sun hours available in your location, your collector area and your hot water load. In many cases, with a household of four, this works out to be around 240L (55gal.) of storage.

Combination space and water heating systems require increased sizes for both collector area (up to 20m²/200 ft²) and storage (500 to 2000L/110 to 440gal.). The proportion of space heating that he system can provide is directly related to the building envelope. A house with a good solar aspect and a well-insulated envelope will have a shortened heating season compared to other houses.

EVACUATED TUBE COLLECTORS.

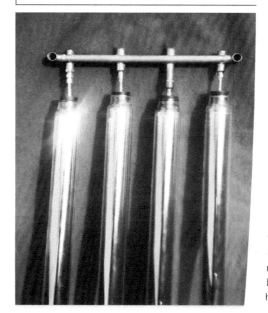

Solar thermal manufacturers are making 'smart' controls for their systems to improve efficiencies through better response times and to better integrate with backup heat systems.

types of system

CLOSED LOOP

A closed loop system is best suited to cold climates, as there is less chance of damage to the collector system due to freezing. The collectors are connected to a heat exchanger by a loop of pipe. The heat exchanger is connected by another loop of pipe to a solar storage tank and a conventional hot water tank that acts as a backup system. The heat exchanger can be stand-alone or integrated into the solar storage tank. The rest of the plumbing and fixtures in the house are conventional, and require no special fittings. A small PV module can be added to power the heat exchanger pump. This is a very good match, as the PV-powered pump only runs when the sun is shining – which happens to be the time that the SDHW system is working.

The collectors absorb heat from the sun, which is then absorbed by the heat transfer liquid that runs in the loop of pipe between the collectors and the heat exchanger. When the collectors become hotter than the water near the bottom of the solar storage tank, an automatic controller switches on a pump. The heat transfer liquid, usually a mixture of non-toxic glycol anti-freeze and water, is pumped from the collectors to the heat exchanger. The heat from the glycol solution is transferred from the heat exchanger to the water in the solar storage tank, then the glycol is pumped back up to the collectors to be heated again.

Photo courtesy of Natural Resources Canada, Building Group.

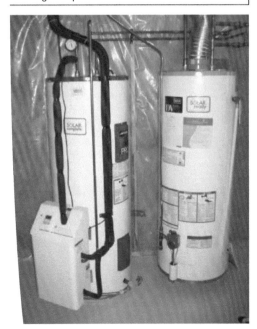

Systems can include a storage tank with an integrated heat exchanger or a stand-alone heat exchanger. Some storage tanks are available with two and three integral heat exchange coils for multiple fuel sources.

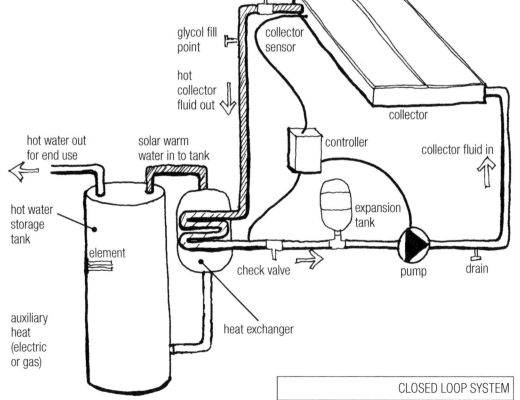

CLOSED LOOP SYSTEM

OPEN LOOP

Open loop systems have similar components to closed loop systems, but the water itself is pumped through the collectors. Cold water is drawn from the bottom of the storage tank and runs through the collectors. The solar-heated water is returned to the middle of the tank, where it rises through convection to the top of the tank to be drawn off as needed. These systems often have problems with corrosion and freezing. They are best suited to seasonal uses such as swimming pools and hot tubs.

THERMOSYPHONING

Thermosyphoning is similar to a convection loop: cool water from the bottom of the storage tank flows to the heat exchanger, where it is heated by the glycol solution returning from the collectors and then rises to the top of the storage tank. The hot water in the solar storage tank is drawn into the backup water heater by city water pressure or the well pump when you turn on the faucet. If your storage tank can be located inside the heated area of the house so that the tank is above the collectors, you can take advantage of thermosyphoning and simplify the system slightly. In this configuration, the storage tank receives heated water coming from the top of the collector into the top of the storage tank. Colder water from the bottom of the storage tank will be drawn into the lower entry of the solar collector to replace the heated water that was thermosyphoned upward. The storage tank may or may not use a heat exchanger.

DRAINBACK/DRAINDOWN

Drainback and draindown systems are also used for SDHW. They both work on the same principle, with minor variations depending on whether you are using a closed or open loop system. Essentially, a sensor in the collectors signals a controller when temperatures drop to about 10°C (40°F). The controller activates a valve that drains the fluid from the collectors and the return/supply piping. When the temperature rises above 10°C (40°F), the controller activates a pump and an air vent to refill the piping and the collectors with fluid, and the valve closes until the next temperature swing. These systems are not suited to most of Canada, as freezing usually causes damage to the more delicate components of the system.

SPATIAL SYSTEMS
Another type of SDHW system is a 'spatial' system. This type of system relies on the heat trapped in a greenhouse or solarium space to passively heat water storage tanks. Typically, they are site-built and are not covered by recognized standards for manufacture or installation. One example of a spatial system would be thin-walled tanks located high in a sunspace to take advantage of the combination of direct solar gain and convective heat currents in the space. As in any type of SDHW system, the amount of heat generated depends on the weather. A certain amount of solar gain will always get translated into heat by spatial SDHW systems, so even on a cloudy day, diffuse radiation is pre-warming the water coming into your hot water delivery system, thus requiring less purchased fuel to bring your water up to the desired temperature.

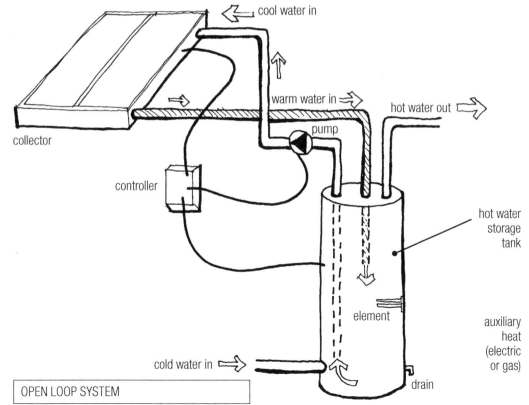

collector

cool water in

warm water in

hot water out

pump

controller

hot water storage tank

element

auxiliary heat (electric or gas)

cold water in

drain

OPEN LOOP SYSTEM

solar hot air systems

Solar air systems provide space heating on sunny days and work on cloudy days as well. There are a few types of systems available. One is a variation on a well-established commercial product that takes the place of siding, called a transpired collector. The other is an 'add-on' collector, similar to a liquid-based glazed collector, that can be mounted on a wall or a roof. A solar air heating unit can help to keep your house cooler in the summer by shading the south wall, and venting the warmed air away from the building to reduce the heating load. You can also bring in cool air by running fans at night. Most air heating systems offer a 10 year warranty, with a life expectancy of at least 25 years, and they are maintenance-free.

TRANSPIRED WALL COLLECTOR.

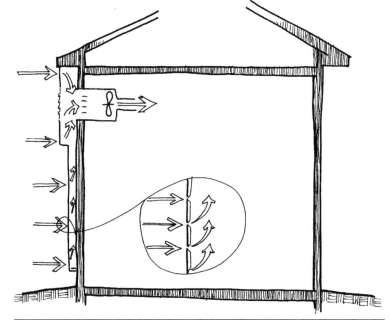

Sizing of transpired wall collector: 1 to 2.5m² per 5L/s (1 to 2ft² for each cfm) of ventilation requirement. The size of collectorvaries from region to region due to the differing levels of solar radiation. This type of system will produce between 300 and 1,000kWh a year for every square metre of collector surface area (28 to 93kWh per square foot, annually). The heat output on a sunny day can be up to 500 W/m²e (150Btu/hr/ft²).

The panels are mounted about 100 mm (4 in) off the weathertight exterior wall with dimensional lumber. A fan pulls air through the panel and blows it into the house. Air can flow directly into the living space or be connected to a forced air heating system for distribution. A differential temperature control turns the solar fan on when the air in the panel is higher than outside.

TRANSPIRED WALL COLLECTORS

This type of system provides low-temperature ventilation air. A metal collector with perforations throughout most of the surface area is attached to an unshaded south wall. As outside air enters the collector and is warmed by the sun it creates a convective current and rises through the collector. When the warmed air reaches the top of the collectors, it enters the building through the home's distribution system, providing fresh pre-heated air to improve indoor air quality.

It is best suited for houses in colder climates where ventilation increases space heating loads. This system is most often used in commercial and large, multi-residential projects because of the look of the metal siding, but has been incorporated successfully into single-family homes in a panelized system. Colour selection is limited to dark blue, dark brown, dark green and black.

FORCED AIR SOLAR HEATERS

Cool air is forced by a fan through a glazed flat-plate collector in this type of system. It can be called a solar forced air heater because the cool air absorbs heat energy generated by the sun and is then forced into your house. Some forced air systems use outside air, others recirculate the cooled air from inside the house through the panel. In a well-designed system, the solar energy air conversion to heat is instant and constant as long as the sun shines – even with light cloud cover.

The collector is similar to the glazed flat plate collectors used in solar hot water systems. A selective glazing maximizes the amount of solar gain, and a special flat black coating on the inside of the box maximizes the absorption of the sun's energy. They operate typically at 70 to 80 per cent efficiency, depending on the make and model, an internal thermostat turns the fan on when the collector reaches a temperature of 100°F or so, and turns the fan off when the collector temperature goes below 70°F or so.

These collectors are modular in nature. Some are designed to sit separately and heat a single room, while others can be linked together to heat larger spaces. Some can be placed in a window, while most models are built to sit on a wall or roof.
Outside fresh air driven through the collectors can also provide pre-heated mechanical ventilation, although adequate filtration of pollutants needs to be addressed. This type of system can also be vented to the outside to provide summer cooling.

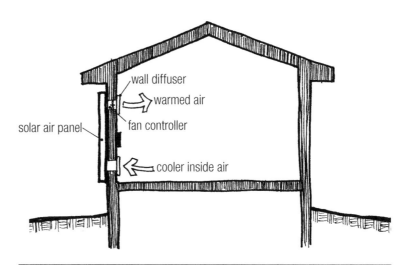

WALL PANELS ARE MOUNTED ABOUT 100mm (4in) OFF THE WEATHERTIGHT EXTERIOR WALL

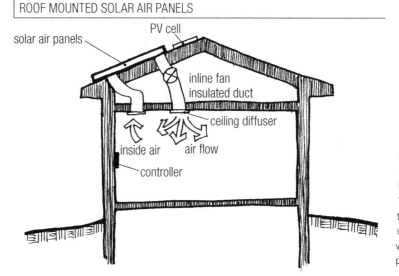

ROOF MOUNTED SOLAR AIR PANELS

These collectors are manufactured in different sizes and shapes and each manufacturer presents the information differently. You can expect to get somewhere between 650 and 1300W per m² (200 and 400 Btu per ft²) at peak capacity. The only operating expense is the cost to power a small fan, which could be eliminated with the addition of a small PV panel.

143

photovoltaics

Photovoltaic (PV) cells are semi-conductor devices that convert sunlight directly into electricity. PV systems offer a number of advantages. They need no fuel, are extremely reliable (when the sun is shining), require little maintenance, are relatively easy to handle and install, are virtually noiseless, virtually pollution-free in operation and have long useable lifespans (over 20 years). Their modular nature means that they can be used for the smallest as well as for the largest applications, and can be added to or subtracted from as the need arises. The amount of power a PV array can deliver depends on the individual cell characteristics, the angle at which the array is installed relative to the sun, the amount of possible solar gain and the area available for the installation.

Conventional PV modules are mounted on a rooftop or in an unshaded area. Manufacturers have developed several different ways to integrate PV into building materials. There are PV roof shingles as well as metal roof sheets with PV embedded in them. PV cells are also bonded to glass to use as skylights, EV roofing and atrium roofs.

Top: Two ways to integrate solar thermal and PV systems.
Left: Solar Shed, Right: roof-mounted.

Photos courtesy of SHIP Database.

Bottom: Roof mounted PV system.

Photo courtesy of Blue Moon Enterprises.

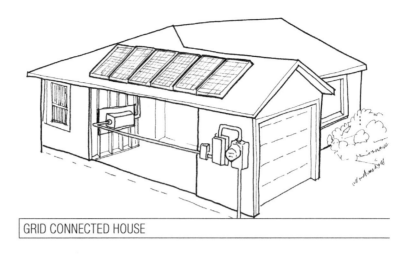

GRID CONNECTED HOUSE

TYPES OF SYSTEMS

Residential systems fall into two broad categories, off-grid, (or stand-alone) and grid-connected.

An off-grid system is not connected to the local utility. All electricity is generated on site and stored for use. Quite often, off-grid homes are in locations where it would cost thousands of dollars to bring electricity to the site, so the owners choose to make the electricity on their own for a similar or lower cost.

To avoid variations in off-grid power, arrays are usually used to charge batteries, which then provide electricity. The most important characteristic of a battery is its storage capacity. Determining the proper amount of storage for an off-grid system is one of the most important parts of its design. An off-grid system will have a few more components than a grid-connected system, including a low-voltage detector to determine if batteries need additional charging.

Auxiliary power is usually used as backup for prolonged periods of overcast skies, and is also required to periodically return your batteries to their fully charged condition (this extends their useable lifespan). Sizing your auxiliary power source requires an in-depth analysis of your estimated electrical load and the amount of electricity your PV array will likely generate. The most commonly used auxiliary power sources are gas or diesel generators.

A grid-connected system creates electricity and feeds its excess power into the utility grid. This eliminates buying and maintaining a battery bank. The grid functions as a limitless battery bank for times when PV output is less than load consumption. When PV output exceeds load consumption, your system feeds back into the grid, giving you a credit. You can still use battery banks to provide backup power when the grid goes down, but they are not required.

In some jurisdictions, grid connected systems require two electrical meters – one meter to record the power into the house and the other to record the power into the grid. Check with your local utility for regulations pertaining to grid-connection and be in contact with them throughout your planning process to ensure you will be able to connect without any surprise costs.

The current cost of PV technology is much higher than the cost of conventional energy in some regions, so grid-connected systems are rarely seen as 'cost-effective'. However, many of our choices in life are not limited by the cost-effectiveness of a purchase. Choosing a solar electric power system comes down to a personal lifestyle decision – just like the type of house or car you may own.

Contact your local planning authority for specific information for planning permission.

Grid-connected PV ties in well with energy efficiency programmes, as homeowners can see how much energy is being produced and how much energy their house is requiring at any given time. Energy monitoring by itself has been shown to drop household energy usage by 10 to 25 per cent.

TYPES OF PV CELLS

There are three types of silicon used in PV modules: single crystal, multicrystalline and amorphous silicon. The most common, single-crystal, is also known as 'mono-crystalline' and written as mono-Si. Extremely thin wafers of silicon are cut from a single crystal, wired together and attached to a module substrate. These are the most efficient silicon cells and have a life expectancy exceeding 25 years. Multicrystalline silicon cells are also extremely thin wafers of silicon but are cut from multiple crystals grown together in an ingot. They are similar to single crystal cells in life expectancy and fragility, but are less efficient and so require more surface area than single crystal cells to produce a given amount of electricity. Thin-film or amorphous silicon PV is made by depositing silicon and other chemicals directly on a substrate such as glass or flexible stainless steel. Thin-film PV materials can look almost like tinted glass. They can be designed to generate electricity while still allowing some light to pass through for daylighting and view. Thin films may be lower cost per square foot, but they have lower efficiencies and produce less electricity per square foot than single-crystal PVs. Several new types of cells with increased conversion efficiencies are being developed, including a liquid-electrolyte cell that does double-duty as a window. Concentrator systems, which focus the equivalent of 50 suns on each solar cell – use inexpensive optics to reduce the number of cells used. Concentrators also minimize the size and number of panels required, but may be better suited to large installations.

Where PV cells are wired in series, the voltages add and the current through each cell is the same. Where they are wired in parallel, the voltage stays the same as for an individual cell, but the currents add. A module is a set of PV cells wired together and held in a frame. Modules are factory-built with various voltage and current levels. To size your system, you must determine the voltage required, then decide on a type of module that meets your needs and calculate the number of modules wired in parallel to produce the amount of current or power you need. A bank of modules wired in parallel is called an array.

PV modules

PV modules do not require heat to produce electricity, in fact, high temperatures actually increase resistance and reduce the voltage within the silicon cells. So, cold climates can actually benefit from modules with lower voltage because more of the power produced is available as charging current, rather than voltage.

It starts with the PV cell, made with special semiconductor materials that allow electrons to be freed from their atoms when the material is exposed to sunlight. Once freed, the electrons move through the material carrying an electric current. The current flows in one direction (like a battery), in other words, direct current (DC). The current generated is directly proportional to the intensity of the sunlight striking the cell.

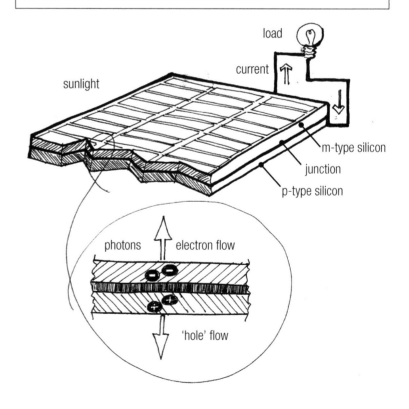

CUTAWAY OF A PHOTOVOLTAIC PANEL

load

current

sunlight

m-type silicon

junction

p-type silicon

photons   electron flow

'hole' flow

performance of PV modules

PV modules generate electricity according to the amount of sunlight falling on them. They produce their rated power in 'peak' sun conditions, that is, when there is full overhead sun – not a wisp of a cloud or a shadow between the sun and the module. Reduced sunlight caused by clouds or shading around the module will diminish the amount of electricity generated. However, modules will produce electricity even when there is no direct sunlight. A cloudy sky with an occasional blue patch will often be equivalent to approximately 50 per cent peak sun, while a cloudy day with rain in the forecast will be about 10 to 20 per cent peak sun. Care must be taken when placing the modules, as shading even one cell of a module will reduce the output of the entire module. Some modules have built in diodes between the cells to reduce this effect, but it is better to mount the solar modules so shading is avoided.

LIFE OF PV MODULES

The estimated lifetime of a PV module is 30 years. Monitoring shows that modules still produce over 80 per cent of the initial power rating after 25 years of service, which makes PV a very reliable technology with low maintenance costs and easy installation. Current and upcoming standards for manufacture and installation set a high level of quality, guaranteeing reliable products.

The energy payback – the time it takes for the PV module to produce as much energy as was used to manufacture it – is less than three years. Over the lifespan of the module it will produce up to 18 times more energy than was used to manufacture it. At the end of the useable lifespan, PV modules can be recycled – silicon, glass, aluminium are all reused. More information is available at www.pvcycle.org.

PV modules are rated at their peak power point. Rated power is a measurement of ideal performance and is based on controlled testing under laboratory conditions. Rated power is measured on a curve showing voltage and current. It is the point where the panel will produce the maximum power in watts. Note: PV modules seldom operate at their peak power point.

In cool climates, look for low voltage (14.5V) and medium voltage (16V) modules. Low voltage modules are good for year round use, and for smaller systems. They may not need a regulator. Medium voltage modules are also good for year-round use in cold climates. They have better summertime battery charging performance than low voltage modules, and are, in most cases, the better choice for household systems, especially those with long wire runs that experience significant voltage loss.

balance of system

There are several other components of a PV system, known collectively as the 'balance of system' (BOS). Typically, there is an inverter, a controller and for an off-grid system, batteries and a charge controller or regulator to protect your expensive batteries from overcharging (which results in water loss and decreased battery life).

A PV array supplies power as direct current (DC). Most household uses require alternating current (AC). For this, your PV system requires an inverter, which will cause a drop in the energy transferred. The amount of energy lost depends on the inverter used. As an example, if an appliance uses 100 Watts of AC power, which it gets from an 80 per cent efficient inverter, then the actual power drain from the system is 125 Watts (100W/0.80 = 125).

Initially, inverters converted power using a 'square wave', which gave terrible fluctuations in the quality of the AC power. The second generation of inverters used a 'modified sine wave', giving better quality power. True sine wave inverters represent the latest inverter technology. The waveform produced by these inverters is equal to, or better than, the power delivered by the utility. All AC appliances operate properly with this type of inverter. They are, however, significantly more expensive than modified sine wave units.

Grid intertie inverters are designed to interact with the utility. They come in two flavours: those that have no battery storage, feeding power directly into the grid, and those that allow your system to function as if it were battery-based, with excess power sold back to the utility. A typical household system will have one large inverter that can synchronize the output with the grid and the household meter.

Many inverters feature a built in charger, so the batteries can be charged from an AC power source such as a generator or the utility. The size of the battery charger is a key feature of the inverter in a back-up power system. If your charger is too small, your generator will run for long periods under a low load to achieve a proper charge, lowering the efficiency of your generator.

Batteries are the heart of an off-grid PV system, storing electricity for use when there is no sun. They also provide a reserve of available energy to run loads that require more power than that provided by your array. Choosing the right batteries, and creating an optimal balance between the size of the PV array and the size of the storage bank is key to getting the best performance out of your system.

What's the best battery for your household system? Well, let's start with the ones that are not suitable. RV and marine batteries might have a label that says 'deep cycle', but they are not adequate for a household and will wear out quickly. Stationary batteries, used by telephone companies for back-up power supply systems, are not designed for deep cycling. Gel cells, filled with an electrolyte, have a much shorter life and are more expensive, but they require no maintenance, tolerate low temperatures, do not spill and do not produce corrosive gases. They are good at remote sites where maintenance is not possible and cold weather prevails.

SO WHAT'S LEFT?

Motive batteries are deep-cycle batteries used to provide energy for electric vehicles such as golf carts and forklifts. They have thick plates that will withstand many deep discharge cycles. These are used for most off-grid systems as they are durable, have good storage capacity and are cost effective. These are usually in 2, 6 or 12V casings. Typically, the golf cart batteries will last 4 to 7 years, while forklift batteries will last 10 to 20 years.

Inverters can now achieve efficiencies as high as 95 per cent. Typically, today's inverters operate above 80 per cent at full power, but the overall efficiency depends on the model and the relative load on the inverter. Although true sine wave inverters are often less efficient than their modified sine wave cousins, the overall efficiency may be comparable because of the energy lost in the harmonics (distortion) of a modified sine wave, especially when running motor loads.

Don't forget that batteries wear out and must eventually be replaced. The deeper they are discharged, the shorter their lifespan. Regular monitoring and maintenance will extend the life span of your battery bank and save you money. Batteries contain toxic materials and should be disposed of properly.

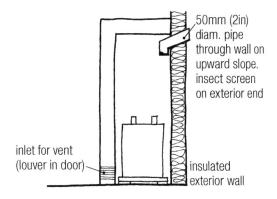

BATTERIES

VENTED BATTERY ENCLOSURE

50mm (2in) diam. pipe through wall on upward slope. insect screen on exterior end

inlet for vent (louver in door)

insulated exterior wall

sizing a PV system

When considering a PV system, the cost-size ratio is usually one of the biggest concerns. Household systems can range from 1kW to 10 kW peak installed capacity. It is always cheaper to save a Watt than make a Watt, so the first order of business is to reduce the required peak capacity of your system by decreasing your electrical load.

Eliminate and replace or minimize appliances that rely on electrically-generated thermal resistance for heat – electric furnaces, water heaters, stoves or ranges, toasters, hair dryers, curling irons, coffee makers and bread makers. All major appliances should be high-efficiency models. Make the most of natural light to reduce the electrical lighting load, and install energy efficient lighting. Fans and other seasonal equipment can be run off small PV cells.

The biggest operational loads in your home, besides any power tools you might use, will typically be the refrigerator and the well pump, if you are on an unserviced building lot. Well pumps that run on DC are available. Changing the pump from AC to DC will reduce the load significantly; however, it may be more expensive than the standard issue AC unit.

Excellent interactive worksheets for sizing your PV system can be found on several websites. A very good place to start is with RETScreen, an online tool for determining the feasibility of renewable energy systems. This free software is available through Natural Resources Canada at www.retscreen.net/

Here is an example of how to gauge the size and cost of a PV system using mono-Si modules.

First off, column one in the table below shows three ranges of potential energy production per square metre/square foot. The actual potential energy production of your system will depend on orientation, the angle at which the modules are to be installed, potential shading and snow cover in your area, as well as seasonal operating temperatures. However, you can use this exercise to set the ground for your system design.

Determine your annual electrical load (see Appendix E to do this). The examples in the table below assume the load to be 3650 kWh/yr (10 kWh/day).

Divide your total annual energy load by the potential production (per m² or ft²) in Column One. This gives you the possible AREA required for your PV array. From this you will know if you have enough existing or planned south-facing roof area to install a PV array that fits your needs, and the number of mounting racks you will need.

Multiply your system area by 0.13 kW/m², an acceptable capacity for PV modules (you might end up with one that is different). This gives you the peak CAPACITY required for your PV array, allowing you to ask around for prices on a system.

Assumed PV system characteristics: PV energy absorption is 100 per cent, nominal efficiency is 13 per cent, average inverter efficiency is 90 per cent.

Location	PV potential annual production		Area required for 3650 kW·h/yr		Array Capacity
	kW·h/m²	kW·h/ft²	m²	ft²	kWpeak
North of 60°	106 to 145	9.9 to 13.5	25 to 34	269 to 365	3.3 to 4.4
Cloudy, mid-latitude	116 to 151	10.8 to 14	24 to 31	258 to 334	3.1 to 4.0
Clear, mid-latitude	137 to 186	12.7 to 17.3	20 to 27	215 to 291	2.6 to 3.5

MAKING IT HAPPEN

Now, your dreams are on paper.

How do you make them real?

If your objectives and expectations are practical, then you have every chance of fulfilling your dreams. It is helpful to ask yourself if you are being realistic about your household? Or are you seeing yourself doing things and living a lifestyle that in reality won't happen?

For the most part, we rely on goods and services offered by others in the construction phase. So, your house will not only have to satisfy your needs, but must also be within the scope of the abilities and experience of the people you are going to rely on to make it happen. Finally, you have to be realistic about costs.

You will find it easier to enjoy the whole process if you can be flexible in your expectations and allow yourself room to react as you go. There is no point in being disappointed when the laws of physics and the rest of the world won't run the way you think they should. If you want to realize your house, it is likely you will make compromises along the way.

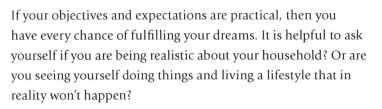

In the beginning, you had some idea of what you wanted for a house. What did it cost when you first imagined it? Was it site-built or factory-built? Who built it? How does the cost vary with size?

costing your project

Before you start to design your house, you may wonder what you can afford. Your costs depend on how your house is to be built and who is building it. Your choices about size, materials, equipment, design and complexity are only some of the variables that affect your costs. How do you arrive at the point where you can gauge the cost of your design? And, how do you avoid wasting a lot of time and energy developing an unaffordable design? Somehow, amongst all of these variables, it is helpful to establish costing guidelines.

Once you have a preliminary design, you may still be unsure how you are going to get it built, or to what standard. But after deciding on the size and most of the materials, it's time to do a preliminary costing.

The best guess is a detailed material list. This involves measuring each and every material to calculate the total quantities required. These quantities are then priced to show how much each of the materials will cost at current market values. This list will be rather extensive and will include all details down to the nails, hardware and finishes. Individual items such as doors, windows and skylights must be priced individually. Take as much time as you need to make sure the list is accurate and inclusive.

A visit to a building supplier with your list of materials will usually get you your prices. If you are not sure which materials you will use, get prices on your options to help you decide. Go to a few suppliers of big-ticket, long-lasting items like windows. Get up close and personal with the windows – do you like how they are finished? If you don't like the butt-welded seams on a vinyl framed unit, but it's the only one that falls into your price range at that showroom, think about how you will feel looking at those seams for the next 25 to 30 years. Go to another supplier to see if they have a differently finished product. Same goes for siding, roofing, trim packages. If you're going to be looking at it, make sure you like it!

With your quantities and prices it is easy to cost the materials for your house. This procedure is necessary especially if you are going to self-build or self-contract. Even if you hire a contractor, it helps to know the material costs.

If you find that your material costs are much higher than you planned, take stock of your facts again. Are some of your objectives unrealistic for your budget? Your home is a major investment, so put as much money as possible into the structure and envelope. Kitchen cabinets, finishes and flooring can be changed without too much extra cost, but increasing insulation levels and improving windows is a costly endeavour. Hot tubs, stainless steel appliances, even solar thermal systems, can all be added after a home is built, as you can afford to pay for them.

There are a number of cost-information sources. For equipment and materials, you can write away or look online for brochures and prices from suppliers. You can visit local dealers and distributors. If you have a builder in mind, ask for references and talk to the owners of houses the builder has recently completed. Some experienced contractors will give you rough costs, usually on a 'per square foot' basis. Check the real estate ads and listings for comparative prices. The figures you get from these sources will guide you in evolving your design.

materials	unit price	quantity	cost
excavation			
stone			
drainage			
backfilling			

Once you have the material costs, you can reassess the realism of your original objectives, If necessary, change them, redesign and revisit the costs. You may have to repeat this cycle a few times before you achieve a workable, satisfying design. Remember, it's much cheaper to change things on paper or on your computer screen than it is to tear down walls or jackhammer concrete!

Maybe your material costs are only a little too high. In this case, you may only have to tighten up the design to bring them within limits. Most people have a fear of things being too small,

SAVING MONEY ON CONSTRUCTION COSTS

There are a few strategies that can be employed to reduce construction costs. One is 'optimal value engineering' (OVE), also known as advanced framing. This revolves around modular units of 2ft: load-bearing studs, floor joists, roof trusses are all lined up on 24in centres, and full sheets of plywood or other sheet materials are used as much as possible. Single top plates, windows placed within stud cavities and other strategies can reduce framing requirements by up to 30 per cent. There are some good resources out there for learning how to use this technique. Some publications are listed in Appendix H. Keywords for internet searches would be 'optimal value engineering', and 'advanced framing'.

Another strategy for reducing construction costs is to get prices from housing manufacturers. You can go to a variety of pre-built systems from panelized construction (also known as structural insulated panels, or SIPs) to a house pre-built in a factory and assembled in large sections on your property. These systems can offer significant savings – up to 30 per cent compared to a site-built house, depending on the complexity of your design. Savings come from several aspects: the factory offers a secure site where weather is not a factor, manufacturers buy materials in bulk and they have assembly lines set up to reduce labour costs. Make sure that you are able to get the best energy-efficient package possible. The foundation work may be up to you. It needs to be well-insulated with good drainage.

so they add a few feet, just to be on the safe side. You have to trust your planning. If your furniture cutouts tell you there is enough room, check it, then believe it.

If, after viewing your objectives and your design, you find that you still do not have your material costs in line, you will have to look at changing your details and materials. Here you have to be careful, replacing one material with another of lesser quality or less desirable characteristics may lead to problems and not pay off in the long run.

There are other ways to approach building affordably, but these depend on your time frame. Do you have the opportunity to build in affordable stages? Some people are comfortable with living on-site in a temporary house, such as a mini-home, while they are building. For some folks, this is simply not an option, but sharing or renting a house might work until you can move into your new house.

The more you spread out your costs (without taking a mortgage), the more easily you will be able to afford to build the house you want. Although there are certain standards that

Look at the items you will be paying for over the 25 to 40 years of your mortgage. Carpet, for example, will have worn out and been replaced once, twice, maybe even three times over that period. Don't sacrifice energy efficiency for short-term savings – you won't be happy when the energy bills come in!

In general, there is not much you can do to change prices of the various materials. But there are strategies you can use to control the cost of your house. We have stressed that design is not a linear process. As you obtain more information, at whatever stage, it should be worked back into the design.

Principal, Interest, Taxes, and Energy (PITE) represent the major costs to homeowners, particularly in the first few years, when maintenance costs should be low. The dollars you spend on energy are just as real as those spent on building or renovating. The money you save by building an energy efficient home can be enough to pay any additional mortgage costs. In fact, energy cost savings continue well beyond the term of typical mortgages, and will increase over the life of a house.

CAN I AFFORD TO BUILD AN ENERGY EFFICIENT HOME?

Building an energy efficient solar house will result in improved long-term resale value and affordability. As energy costs increase, the value of houses with minimal operating costs will go up. Some lending institutions already give higher market value and reduced rates on mortgages and building loans for energy efficient houses.

In the US, Energy Efficient Mortgages (EEMs) for new houses and Energy Improvement Mortgages (EIMs) for existing houses, stretch debt-to-income ratios, allowing more financing based on a home energy rating. The Energy Star website has more details.

In the UK and other countries, low interest green mortgages are also available to fund the installation of solar panels, high-performance windows and other energy saving devices.

Other financing options include municipal or council bond programmes, where the upgrade cost are 'attached' to the property tax for the house at zero or low interest rates over a long period. This kind of financing option makes long-term investments in energy efficiency and renewable energy more palatable for many homeowners.

must be met before you can obtain an occupancy permit, you can opt to do your finishing in stages. If you are willing to live on painted plywood floors for a few years, then you can invest the mortgage money that was destined to go into car-petting in a better structure, more energy efficient windows or a solar domestic hot water system. To sum up, you can control your costs by changing your requirements, your design, your materials and/or how you build.

WHAT'S IT WORTH?

In the UK, the Committee on Climate Change calculates that the owner of a typical three-bedroom Victorian end-of-terrace home, with three exposed walls, could expect to spend £10,280 on a package of energy-saving measures which could include replacing the wooden frame windows with energy efficient UPVC windows and a new boiler. (from an article entitled: 'UK to Offer Cheap Green Mortgages to Foster Energy-Saving Homes', Jonathan Leake, Sunday Times, 12 July 2009.

This would cost the householder £514 a year in repayments over 20 years, assuming a zero interest rate through a council tax-reduction arrangement, while the savings on energy would total £802, a yearly 'profit' of £288.

In the US, let's say a modest new rancher requires a mortgage of $150,000. A standard mortgage might be 7 per cent over 25 years. Upgrading to a more energy efficient building envelope and mechanical system might see an additional 10 per cent in costs. In addition to a stretch in debt-to-income qualifying ratios through an EEM, you may also expect to see a drop in mortgage rate. So let's say the 'green' mortgage ends up being $165,000 at 5 per cent over 25 years. This increases a monthly payment by $90, while the decrease in energy costs could range from $100 to $175 a month, depending on where you live, a yearly 'profit' of anywhere between $120 and $1020. The calculations above do not allow for increases in the cost of energy.

Photo courtesy of Abri Sustainable Design

getting it built

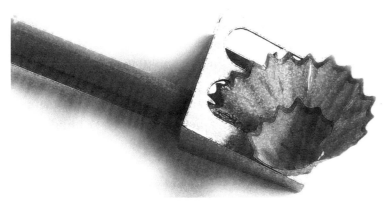

Once your costs are under control, how do you get it built? You can hire a general contractor to oversee the whole process, you can act as your own general contractor, or you can build it yourself.

Hiring a general contractor means that person gets all permits, arranges everything and supervises the work of all the subcontractors and tradespeople through to completion. You pay the bill, or yell and scream if it is not done to your satisfaction. Many contractors are part of a builder's association and can offer you a warranty on their work.

The onus is on you to ask questions and get references. It is not enough that the contractor intends to do good work. You have to be sure that they are capable. Similarly, the contractor should feel sure that you are capable of carrying out your end of the agreement – to pay. If you feel that the contractor's price is too high, you can ask for a breakdown and an explanation. If you are still not sure, get several other estimates. The investment is large enough to warrant your being careful. However, don't overdo it. The key is TRUST – trust founded on explicit information, clear understanding, competency and good will. You need to be comfortable with the person who is in charge of building your house and spending your money.

CONTRACTS

You need a construction contract. You need a legal agreement that gives you protection from poorly done or incomplete work. This also protects contractors, as they can outline exactly what they are responsible for, and what they are not responsible for.

Your income or savings must cover not only your regular living expenses, but the monthly interest on your construction loan as well. If you expect to borrow money from a bank, your credit rating must be in good shape. You will probably have to scrape up considerable cash to tide you over between bank payouts. You should have adequate savings or a line of credit for unforeseen expenses and emergencies.

Building a house really is a full-time job. Save up as much vacation time as possible. If you have to work at another job while you build, it should be a flexible one that allows you to take time off when you need it. If your spouse also has a job, stagger your vacations so that one of you is always around to supervise construction. Be prepared to sacrifice most of your free time to complete the house on schedule. Even if you don't personally have a tight schedule, the bank or building department might force one on you.

You need a sense of humour. House construction is chaotic. Problems will arise and you will be under a lot of pressure. If you stay cool, calm and collected, you will have a very positive effect on the process and everyone who's working with you.

Building your own house can help you lower costs, or get more house for your money. If you are well-organized and have some experience, you can save from 20 to 60 per cent, depending on whether you serve as your own general contractor or provide the labour as well. The single most important trait of a successful owner-builder is organization. Another important requirement for being a successful owner-builder is good communication skills. You should be a good listener and arbitrator. You must be

There are some very good publications available in print or online, that describe the types of contracts typically used in the construction industry, and how to make sure the one offered you suits your project.

RULE OF THUMB FOR KEEPING YOUR SANITY DURING CONSTRUCTION

Whatever it costs, add 15 per cent. Whenever it's supposed to be finished, add three weeks.

able to communicate what you want. You must also be able to hold your temper despite extreme frustrations.

With self-contracting, you become the general contractor. You get the permits, arrange everything and supervise the work. It usually involves a lot of running around, so you have to be very well organized. Also, the people doing the work have to be competent and you have to know that they are. It requires that you know about the construction process. Was it done right? What comes next? Again, you have to make agreements and get estimates.

With the self-build method, you do all of the above, plus some or all of the construction yourself. Again, you are displacing someone else's paid time with your own. The key is to determine very carefully what you are capable of and have the time to do. If you have not built before, or you do not have the time, skill, knowledge or tools, you may be in for some hard-earned experience. How much value do you place on this learning experience?

If you self-contract or self-build, doing it yourself JUST TO SAVE MONEY usually doesn't work. You have to want the experience and you have to be realistic in your self-expectations. In either case, you have a good start if you (or your mate) are handy with tools and enjoy doing repairs around the house. Can you read a set of plans? Can you make your way around a lumberyard? Can you converse with subcontractors, inspectors and suppliers in builder's jargon? Having this knowledge, or some of it, before you start the building process will help.

If you do the bulk of the work, you assume control of the construction. If it goes wrong, point the finger at yourself. If it goes right, take a bow. You're taking a chance, but you control the odds. If you are unsure of yourself, you could hire an experienced carpenter to work with you. But can you afford that?

SOME TIPS FOR MAKING IT HAPPEN
- *Show your plans to friends. Do you know anyone who is a builder or a home designer? If you feel comfortable doing so, ask their opinion of your design. Remember, though, that this is what these people do for a living – offer to pay for their time.*
- *Read as many books, watch as many DVDs, videos and TV shows as possible, Know the nuts and bolts of the whole building process.*
- *Check your subcontractors thoroughly. Some contractors view owner-builders as easy marks. After all, chances are you will never hire them again.*
- *Don't skimp on materials or attention to details. It is tempting to cut corners, but opting for quality throughout the house will be rewarding, especially around such issues as energy efficiency, waterproofing, etc.*

- *Don't give in to mistakes. Redo what is going to make you crazy to look at over the next 20 or 30 years.*
- *Play it safe with insurance. Building materials – especially in the framing stage – are delivered in bulk to the site. Some inconsiderate souls would use your paid-for framing for their designs! Think about third-party liability too: if someone suffers an injury on your site, you could be financially ruined for life.*
- *Keep accurate financial records. A detailed account of every penny is a valuable record for insurance and tax purposes.*
- *Be aware of time management. Physical work gives instant gratification – swinging that hammer shows progress, but your most important function may be to act as project coordinator. Don't run yourself ragged for the sake of a few dollars.*
- *Keep your sanity and your health.*
- *Don't rush to move in.*

scheduling

Your best insurance for success is a well-thought-out construction schedule. The more detailed you can make this schedule, the better the odds for success. You have to imagine doing it step-by-step, what each step entails, how long it is going to take, how messy it is, what disruption it will cause to the family routine, how your household can work as a team, who is doing what, and so on. Plan for things to go wrong, and build in strategies for handling problems.

Activity	Equipment Needed	Materials Needed	Work Done By	Time Required	Dependent On/ Follows What Activity?
Preparation					
site designing					
shelter designing					
getting building permit					
stake out shelter					
clear house site					
check access: trucks and equipment					
arrange excavation and gravel					
Foundation work					
excavation					
gravel placement					
gravel leveled					
arrange for foundation work					
rain, ductwork, plumbing etc. for underslab					
arrange concrete truck and access					
waterproof					
backfill					
Getting Structure Roof Tight					
arrange lumber delivery					
frame walls					
sheathe walls					
frame roof					
sheathe roof					
arrange and install windows and doors					
install house wrap					
apply rainscreen and siding					
trim, paint					
rough landscaping					
Interior Work					

With good planning, you can insure that your energies are well-spent. List everything that has to happen, and put it down in sequence. Break it down into stages and assign time guess-timates. Allow time for rest, and be good to yourself. Include time to marvel at your work and take pictures of your progress. Keep a diary, and get everything out of the experience that you can.

157

CASE STUDIES

Acadian post and beam

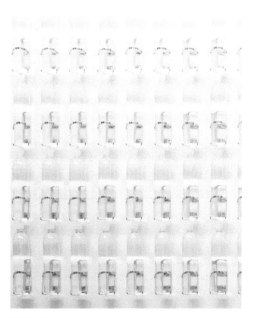

F10 House

PassivHaus

Cape Breton grid-interactive

Ontario off-grid

'Roscoe's rules'

acadian post and beam

LUNENBURG, NOVA SCOTIA 2003

A FEW KEY FACTS

- *Main floor South glazed area: 10.2m² (110ft²)*
- *Heated Space: 160m² (1715ft²)*
  South window to main floor area: 11 per cent
- *Fuel Source Displaced: full-size oil space/water heating system*
- *Estimated GHG Emission Reduction: 9600kg/year of CO_2 equivalent*
- *Approximate Savings: Annual purchased energy less than 5,000kWh-e/year, not including wood use.*
- *Payback: Investing in energy efficient envelope and high-performance windows added ≈10 per cent to general costs*

Designer: Abri Sustainable Design & Consulting
email: shawna@abridesign.com
URL: www.abridesign.com
phone +1.902.489.1014

The frame of this house is from an original Acadian home in Digby County, Nova Scotia that was dismantled and raised on a property in Lunenburg County. As a retirement home, the owners want a house that costs them little or nothing to operate, and can accommodate them and their aging parents, one of whom is wheelchair bound. The house is designed to maximize passive solar access and create a completely barrier-free main floor space.

An extension to the south face of the main floor adds space for the dining room and a portion of the main sitting area. A dormer added onto the south upper floor increases the usable space in the main bedroom. The original look of the house is maintained inside and out by using the original sheathing boards.

This house features: Passive solar design planning for solar thermal space and water heating systems.

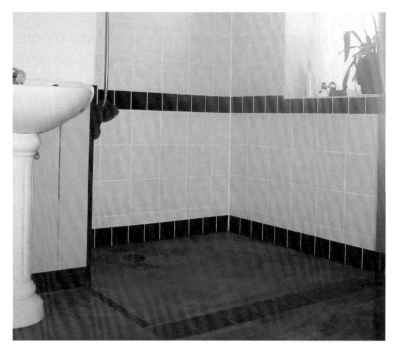

Photos courtesy of Shawna Henderson, Abri Sustainable Design.

To leave the frame exposed on the interior of the house, the exterior is wrapped with rigid foam insulation, with additional batt insulation (RSI 3.5/R20) installed on the North wall of the house. The main roof has a double layer of rigid foam insulation installed (RSI7/R40). The windows are European-style 'tilt and turn', double-glazed, low-e coated with insulating spacers and argon fill, made in Nova Scotia. A tight building envelope means little heat loss through exfiltration. A high efficiency HRV takes care of the ventilation needs.

An EPA-approved woodstove provides backup space heat. The homeowners annually harvest two to four cords of wood from their own woodlot. This house also has two 40gal electric hot water tanks: one for domestic hot water and the other for back-up space heat, in the form of hydronic in-floor heating. The hydronic system was installed in anticipation of the owners being unable to carry out woodcutting work as they age. Should the hydronic system be used as full-time backup in the future, the house is pre-plumbed for a combined solar DHW/space heating system. This house requires about 5000kWh a year of purchased energy, a 70 per cent reduction from the energy required to run a comparable conventionally designed and built house.

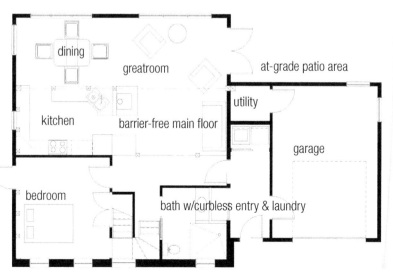

Case Study Sources:

Abri Sustainable Design

Canadian Solar Industries Association (CanSIA)

Canada Mortgage and Housing Corporation (CMHC)

161

f10 house

CHICAGO, ILLINOIS 2003

A FEW KEY FACTS

- *Heated Space: 170.1m² (1830ft²)*
  Solar Chimney augments daylighting as well as heating and cooling regimes.
- *Energy Source: 90 per cent efficient natural gas boiler – grid electric*
- *Annual Space Heating Requirements: 60MMBtu (372MJ/m² or 32.8kBtu/ft²) natural gas*

Designer: Esherick Homsey Dodge & Davis
email: info@ehdd.com
URL: www.ehdd.com
phone +1.312-655-0690

F10 House was one of five affordable case-study designs chosen for the 2000 New Homes for Chicago programme. The house sits on a narrow city lot with an east-west orientation, and is surrounded by adjacent buildings. It was designed to reduce life-cycle environmental impacts by a factor of 10 compared to the average home built in America. Finished in 2003, F10 House was chosen as an AIA Top Ten Green Project for 2004.

The building envelope is super-insulated. A modular design minimized waste and allowed off-site assembly. Materials were selected for their durability and low production impact. The site includes low-maintenance plantings, pavers and a green roof to minimize storm water runoff, making the entire site permeable.

This house features:
passive solar heating
and cooling
solar chimney
daylighting

A solar chimney – a vertical shaft with south-facing operable clerestory glazing – brings light into the centre of the house. The solar chimney also collects heat in the winter, distributing it back into the living space using a high-output ceiling fan.

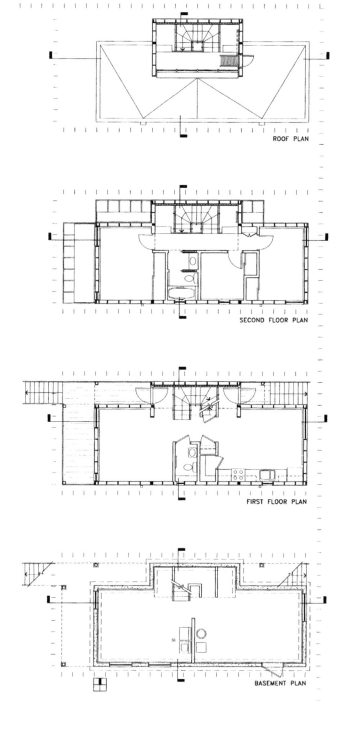

ROOF PLAN

SECOND FLOOR PLAN

FIRST FLOOR PLAN

BASEMENT PLAN

A wall of water bottles acts as a short-term heat sink. Both of these strategies minimize the amount of energy used by the gas-fired baseboard heating system.

In the summer, a whole-house fan 'primes' the solar chimney by drawing hot, stagnant air out of the house when windows are opened. Once the natural draft is running through the house, the fan can be turned off. An open floor plan enhances cross ventilation. The bottle-wall heat sink and green roof also help to reduce heat load in the summer. This is one of the only houses built by the Chicago Department of Housing in recent years that did not have a central air-conditioning system.

Building height and sun angles were calculated to insure that solar insolation would not be blocked by adjacent structures to the south. Sunshades were provided on west-facing windows to minimize heat gain generated from 2:00 to 5:00pm in the summer. Late-afternoon sun is blocked by houses across the street. Heat loss is minimized through window placement. Small rooms were located along the south flank to take advantage of the sun.

Controlling systems during most of the year consists of adjusting a single rheostat to operate the speed of the ceiling fans. The porch includes perforated metal awnings that provide shade and channel air into the house.

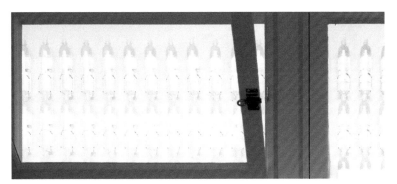

Case Study Sources:

www.ehdd.com

Green Homes for Chicago Publication, available at egov.cityofchicago.org/

Environmental Building News Vol. 10, No.1 p. 2 (01/2001) http://www.buildinggreen.com/auth/article.cfm?fileName=100102a.xml&page=1

passivhaus

LINDLAR-HOHKEPPEL, GERMANY 1998

A Few Key Facts

- *Type: Five single family buildings built around a courtyard*
- *Built: 1998*
- *Characteristics: Solar collectors for domestic hot-water, ventilation system with heat recovery, rainwater harvesting for toilettes and washing machine*
- *U-value exterior wall: 0.105W/(m²K)*
- *PHPP specific annual heating requirement: 10.9kWh/(m²a)*
- *U-value under slab: 0.101W/(m²K)*
- *U-value roof: 0.099W/(m²K)*
- *PHPP annual Primary energy demand 43.5kWh/(m²a)*
- *U-value windows: 0.745W/(m²K)*
- *Heat recovery: 90 per cent*
- *Blower door test (50Pa): 0.4AC/hr*

Designer: Manfred Brausem

The objective of this pilot project was to prove that a passive house can be built at the same price as a comparable sized home, within the same time period, with conventional technologies and without any public subsidies (which were not available at that time). After more than nine years evaluation, the house still fulfils quality, comfort and energy efficiency expectations. Heating energy consumption and costs are about 1/10 of a comparable conventional home.

The north side of the house has lower room heights. Here you will find the carport and storage which create additional weather protection. In the centre of the house is the mechanical room which supplies the kitchen and baths in the upper floor through a supply chase. On the south side of the building are the living

This house features:
- passive solar design
- solar thermal space and water heating
- grid-interactive PV

164

areas – dining, living and bedrooms – which have increased room heights. These rooms can be separated and can be adapted to individual uses, e.g., sleeping loft. The cross section also shows the sun angles in different seasons.

The windows act like passive-solar 'collectors' and earn real, solar energy gains: Nevertheless, the U values of the windows must be less than 0.8W/m²K. Special, high-insulated window frames with triple-glazing are necessary.

The air delivery system was planned with short duct runs for low flow resistance and was installed invisibly within the ceiling beams. The use of silencers avoids telephony effects.

Heat is distributed through the heat recovery in the ventilation system. The HRV unit (heat recovery ventilation unit)

has a cross flow air to air heat exchanger. A mini LP warm air heater which is installed in the bypass of the supply air channel delivers the small amount of the remaining heating demand thermo-statically controlled through the fresh air ducts to the living spaces. The liquid gas is stored in 2 gas cylinders with 33kg of contents outside in the storage shed. The heating costs experienced by the inhabitants, with approximately 4 liquid gas bottles of 33kg per heating period, confirmed the projected calculations.

The domestic hot water demand is being met by more than 60 per cent with a generously dimensioned solar thermal system. An on-demand electric post heater is used as backup only if the situation should occur that the solar energies should not suffice – and only so much, so long and so hot as the user desires for his hot-water needs. Thus the solar gains can

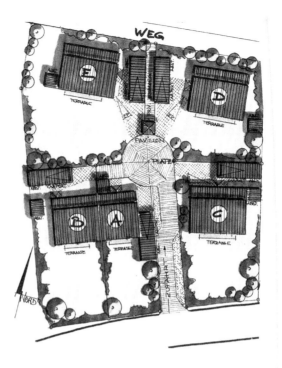

be maximized, the post heating remains automatically switched off if no need exists.

All exterior walls and the roof struture are planned to be vapour diffusion open. Vapour barriers were completely avoided, because their flawless installation, without any damage is too hard to realize. The airtight plane and vapour barrier is realized in walls and roof by the internal 15mm-OSB sheathing. All panel joints were sealed by tongue and groove and PU adhesive. All abutting joints to building components like slab, windows were sealed with especially approved adhesive tapes.

The air tightness value for all the passive houses was under 0.5AC/hr at 50Pa of pressure.

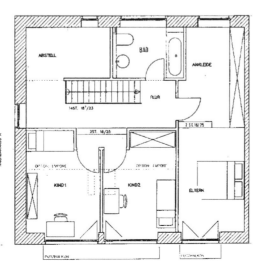

grid-interactive straw-bale

MISSISSAUGA, ONTARIO 2000

A Few Key Facts

- *PV System Size: 2.3kWp capacity, 18 UniSolar Standing Seam SSR-128 Roof Panels (amorphous)*
- *Inverter: Trace SW 4000*
- *Net metering equipment: Standard Residential kWh Meter*
- *Average annual energy production: Residents can read electricity generated on two battery regulator digital displays, system can deliver between 2270 and 2850kWh annually.*
- *Solar Thermal System: 6m² (65ft²) flat-plate collector with 240L (50gal) storage tank, supplies approximately 10GJ (2800kWh-e) energy*
- *Fuel Source Displaced: grid electricity and gas-fired hot water system*
- *Estimated GHG Emission Reduction: 700 to 1000kg/year of CO_2 equivalent*
- *Approximate Savings: 55 to 61 per cent of fuel for hot water, 30 to 40 per cent of electricity use*
- *Payback: PV system cost $28,000 in 2000. Payback was not one of the motivating factors to install the system. Payback on the solar thermal system would be between four and eight years.*

Designer: Sol Source Engineering
email: perdrewes@rogers.com
phone: +1.905.898.0098

This house features:
solar thermal water heating
grid-tied PV with battery
storage

This house showcases several innovative approaches to energy efficiency and energy production, including straw bale walls (≈RSI-8/R-40) and standing seam metal roofing, one quarter of which is a photovoltaic array. The owners want the house to reflect a commitment to sustainable living. In addition to the utility interactive PV system, which also has stand-by power capability, the house features solar thermal water heating, energy efficient appliances and radiant floor heating. The house is finished with allergy-free materials. The owners' choice of location – a suburban area with strong ratepayer and planning controls – was based partially on the availability of pub-lic transportation, in keeping with their desire to reduce their personal impact on the environment. As such, the project had to conform to regulations related to orientation to the street, resulting in some loss in PV performance.

Although the utility, (Mississauga Hydro) had no official grid-connect policy in 2000, they were informed of the project at the outset. After inspection and approval by the Ontario Electrical Inspection Authority (OESA), the utility gave verbal permission to connect. Mississauga Hydro had no official net metering policy when this project was carried out, so there were no changes to normal billing procedures.

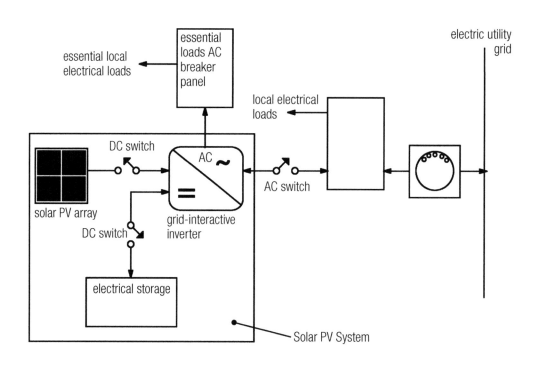

The inspection authority had some concern about battery venting requirements but the installation of wiring in the straw bale walls was more of an issue. The integration of the PV system in the roof required careful coordination with other building trades, especially the roof installers. Because the system has stand-by power capability, a separate sub-panel was provided for critical loads. Deciding which were critical loads and providing separate feeds to these loads required specific instructions to the electrician. However, most tradespeople took an interest in being part of a new and innovative project.

Photos courtesy of Per Drewes, Sol Source Engineering.

Case Study Sources:

Sol Source Engineering

Canadian Solar Industries Association (CanSIA)

International Energy Agency (IEA)

ontario off-grid

ORANGEVILLE, ONTARIO 1998

A FEW KEY FACTS

- *Main floor South glazed area: 16.7m² (180ft²)*
- *Heated Space: 149m² (1600ft²)*
- *South window to main floor area: 11 per cent*
- *PV system size: 400Wp*
- *Wind turbine: 400Wp*
- *Energy storage: 8–6V (225Amp) golf cart batteries*
- *Fuel Source Displaced: full-size natural gas space/water heating system, grid electricity (generation mix = nuclear, hydro, coal)*
- *Estimated GHG Emission Reduction: virtually eliminated.*
- *Approximate Saving & Payback: This remote location meant high installation costs for grid-connected electricity, on-site generation was more economical*

Designer: Greg Allen

This off-grid house is designed to be ecologically responsible. It is earth-sheltered, superinsulated and heated by passive solar and a wood-fired masonry stove. The envelope is highly efficient, so the inside temperature never falls below 10°C and is often above 18°C, without added heat. This house has zero

greenhouse gas emissions associated with space and water heating: the wood-burning masonry heater is considered 'carbon-neutral' (burning wood efficiently produces no more emissions than if the tree had died and rotted in the forest). The solar thermal system produces no emissions. Nearly all of the electricity is

This house features:
Passive solar design
Masonry heater
Solar thermal water heating
Off-grid photovoltaics

supplied by a small hybrid PV/wind system. The only other emissions related to operating the house come from the propane kitchen range (summer use only) and the gasoline-powered generator used occasionally for backup power.

The earth-bermed design reduces heating and cooling requirements significantly by taking advantage of the constant temperature of the earth and reducing the impact of the winter winds from the north and northwest. The initial site selection faced a northwest view. The site was changed to reduce the impact of the building on the property and to maximize the passive solar potential of the house. Fill from the excavation was used to create a level front yard.

Local materials were used throughout the house. Woodchip-cement blocks with mineral wool insulation make up the bulk of the wall system. Recycled steel posts were included in the structure and reused bricks add thermal mass to the house as finish materials for the masonry heater, the dividing walls and the planters that make up the greywater treatment system. Dry site conditions dictate that extensive water conserving features be built into the house – flat roofs feed a cistern for laundry and bathing to reduce the demand on the spring-fed well. Two composting toilets further reduce water demand. Greywater from laundry and bathing is fed to an indoor planter and out to a small sub-surface wetland system.

UPPER FLOOR

south →

MAIN FLOOR

Photos courtesy of Greg Allen/Cobalt Engineering.

Case Study Sources
Greg Allen

Cobalt Engineering

Canada Housing and Mortgage Corporation (CMHC)

'roscoe's rules'

NOVA SCOTIA

A FEW KEY FACTS

Don Roscoe has been designing high-mass passive solar homes that feature forced air systems and monolithic foundations with 'activated' slabs since the 1970s. This case study looks at his specific approach to passive solar design and climate control. Some houses also feature solar thermal and photovoltaics.
- *South window to main floor area: 10 to 15 per cent*
- *Fuel Source Displaced: full-size oil space/water heating system*
- *Approximate Savings: 50 to 65 per cent space heating*
- *Payback: Investing in passive solar design and high-performance windows has a payback of less than five years*

Designer: Don Roscoe, solar home designer/builder
email: solardon@chebucto.ns.ca
phone +1.902.852.3789

This approach to passive solar design combines many solar and conservation characteristics in an integrated geometry, and can be broken out into ten 'rules'.

1. Stack the living spaces in multi-levels on the south side of the house

2. Locate utility areas on the single-storey, earth-bermed north face

3. Extend the saltbox roof from the grade level on the north to the two-storey south

4. Shape the exterior to create protected outdoor courtyards

5. Open the interior south core to the second-floor ceiling to create an indoor courtyard, overlooked by balconies and interior windows

6. Move the basement from under the house to behind the house (on the north side of the building) and build on a heat storage slab

7. Install a thermal break between the heat storage slab and the concrete under unheated spaces

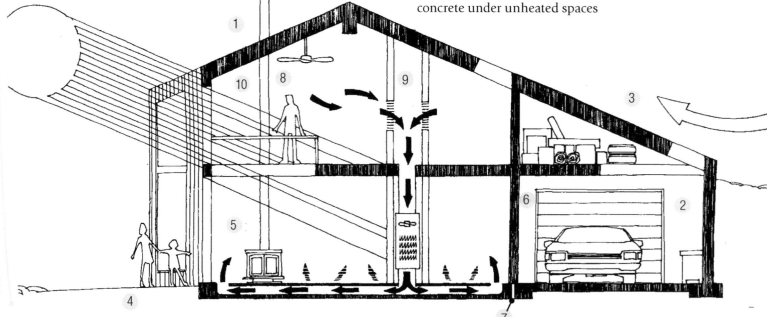

This house features:
Passive solar design

8   Use ceiling fans for cooling as well as even heat distribution in open spaces

9   Use an air recirculation system with a fresh air intake to distribute heat from solar gain or any other source, to filter the air, to charge and discharge the heat storage slab and to assist your ventilation system

10 Install chimneys inside the thermal envelope of the house

The air recirculation system is based on a small electric forced air furnace that supplies the blower unit. The blower runs year round, to heat up and cool down the slab and to keep the temperatures in the house from stratifying. Many of these houses have a small woodstove to augment the passive solar gain during the coldest parts of the winter. The thickened slab, with ductwork throughout, acts as a heat sink, absorbing heat during the day and releasing it at night.

Photos courtesy of Don Roscoe.

Case Study Sources:

Don Roscoe, Solar Designer/ Builder

Harrowsmith Magazine

SHIP Database

APPENDIX

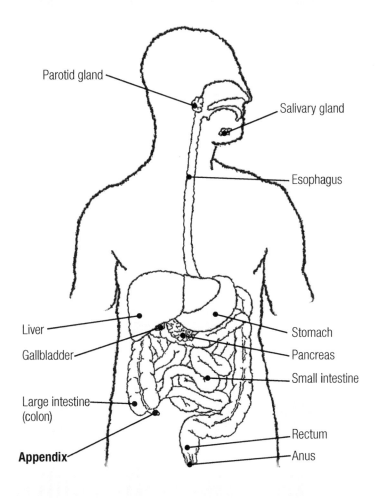

Parotid gland

Salivary gland

Esophagus

Liver

Gallbladder

Large intestine (colon)

Appendix

Stomach

Pancreas

Small intestine

Rectum

Anus

Illustration: Ideas Ink Design

Your appendix is a small, closed tube that is about the size of your finger. It attaches to the beginning of your large intestine, where the small and large intestines meet. It is open at the end that connects to the large intestine and closed at the other end.

It does not appear that the appendix does anything useful. This appendix, however is bursting with useful information.

A. sunpaths & obstructions
B. maps & declination
C. insulation values
D. heatloss calculations
E. sizing PV systems
F. rural site considerations
G. earthworks
H. resources

A. sunpaths and obstructions

To determine the time of day and the time of year when direct sun is blocked from reaching any point on your site, it is necessary to plot the obstructions as seen from that point. This step is unnecessary if the skyline to the south of your site is low and there are no tall trees or buildings near where you plan to set your house. To plot obstructions, we use two tools. One is a sunchart. The other is an altitude measuring device, described below.

Suncharts for three different latitudes are shown on pages 176–184. Please use the most appropriate to your location.

As you can see the sun's path is charted for the 21st day of each month. The sun's path is longest during the summer months when it is highest in the sky, rising and setting at the furthest points away from true south. Just the opposite is true for the winter months.

To plot obstructions, place yourself at the approximate location on the site where you want your house. Facing true south (even if your house will not be facing true south). Aim your measuring device in the direction you are facing, determine the altitude of the skyline and mark the point on the sunchart at the line indicating south.

Determine and record the altitude angle of the skyline every 15° along the horizon. You should have a total of 17 points marked on your sunchart. Connect them with a line.

For obstructions such as trees and buildings that will block the sun during winter, find the altitude for each. If they fall between the 15° increments on the sundial, estimate their position on the horizon. Indicate deciduous trees with a dotted line, as they will not completely obstruct the winter sun (see page 13).

The resulting graph shows the complete skyline. The area below the line shows the times throughout the year when direct sun will be blocked at the point from which you plotted the sunchart.

We have used simple tools and methods throughout this manual to determine if such things as direction and altitude. The degree of accuracy is adequate for most applications; you need more accurate measurements, you will need more sophisticated tools, such as transits and levels. You can use a compass to determine south, but you must be certain to correct for magnetic declination, as well as any nearby metal, which can distort your compass reading.

Mark out 10° increments on a semi-circle of heavy cardboard. Make a hole at the centre point and thread a weighted string through it. This acts as a plumbob.

Always measure the altitude from the same height above the ground (your eye level when standing). The end of the device furthest from your eye should be level with the topmost point of the object you are sighting.

The altitude angle in this example is 30° from the horizon.

To use the device, site along the flat edge of the highest point on the skyline or object and let the plumbob hang straight down. Hold the string firmly (without moving it) and note the resulting angle on the sunchart. The angle is the altitude of the skyline or object.

HOW TO MAKE AN ALTITUDE
MEASURING DEVICE

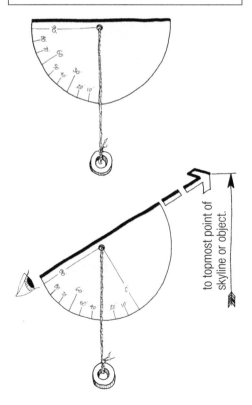

to topmost point of skyline or object.

First plot the skyline, then plot the position of nearby obstructions that may shade your house. Note the angle on the measuring device, then plot each point on the sunchart.

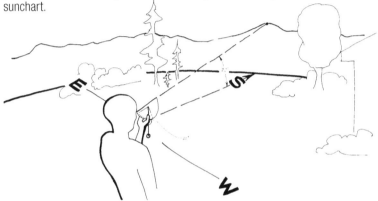

120° off south corresponds to the direction of 4 o'clock and 8 o'clock on the sundial. Remember that the sun moves across the sky at a rate of 15° every hour. *(See page 12 for more on finding south)*.

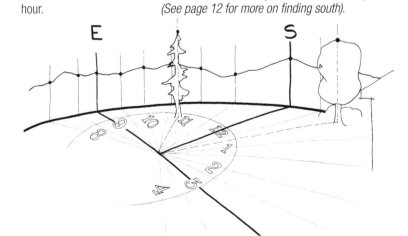

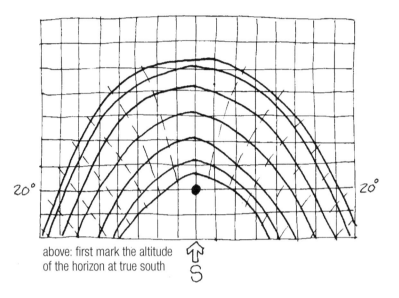

above: first mark the altitude of the horizon at true south ⇧ S

below: chart points to 120° either side of south

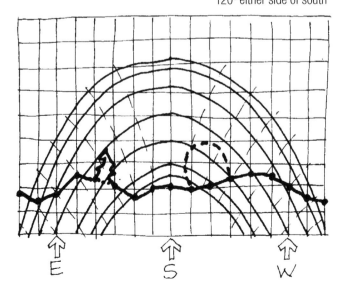

⇧ E ⇧ S ⇧ W

Following are suncharts for different latitudes that will help you define what part of the sky is unobstructed at what time of day throughout the year. (Shown on next few pages).

Sunchart is a vertical projection of the sun's path as seen from the Earth. The grid represents the vertical and horizontal angles of the whole skydome – the visible hemisphere of the sky above the horizon – stretched out and laid flat. The hours of the day are also plotted on the charts (the heavy dotted line that runs through the monthly sun paths). Note that these charts are in standard time, not daylight savings time.

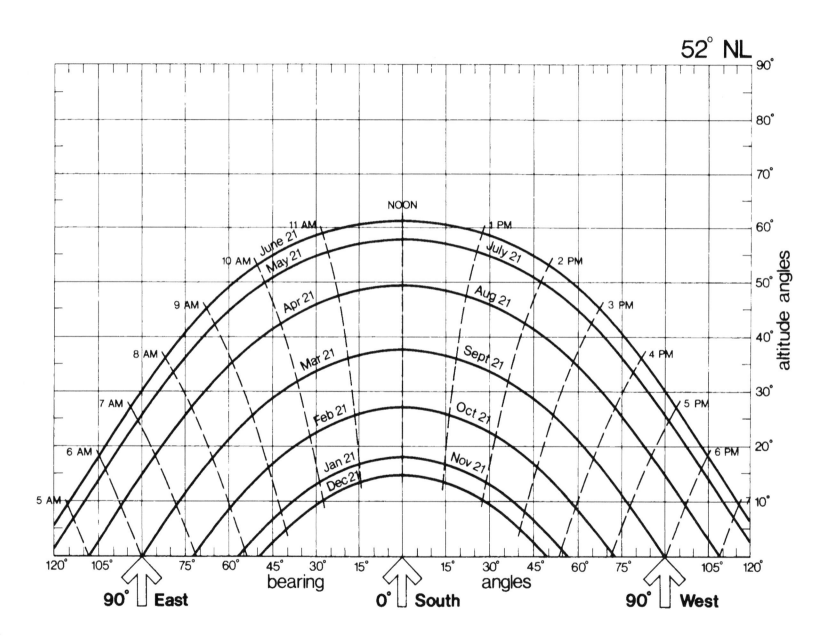

52° NL

altitude angles

90°
80°
70°
60°
50°
40°
30°
20°
10°

NOON

11 AM   1 PM

June 21   July 21

10 AM   May 21   2 PM

9 AM   Apr 21   3 PM

8 AM   Aug 21

7 AM   Mar 21   4 PM

Sept 21   5 PM

6 AM   Feb 21   Oct 21   6 PM

5 AM   Jan 21   Nov 21

Dec 21

120° 105° 75° 60° 45° 30° 15° 15° 30° 45° 60° 75° 105° 120°

bearing   angles

90° ⇧ **East**   **0°** ⇧ **South**   **90°** ⇧ **West**

Suncharts from
The Passive Solar Energy
Book
(Mazria, Pages 319–321)

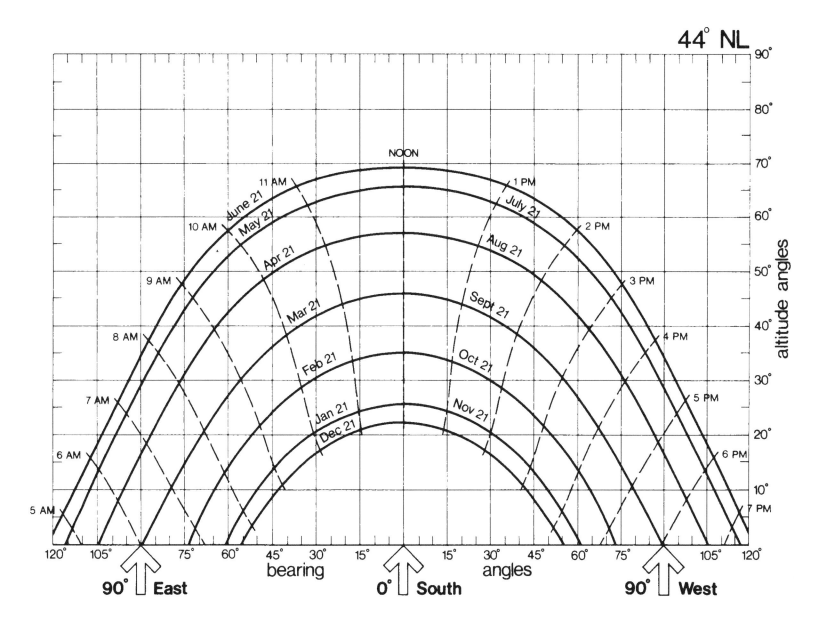

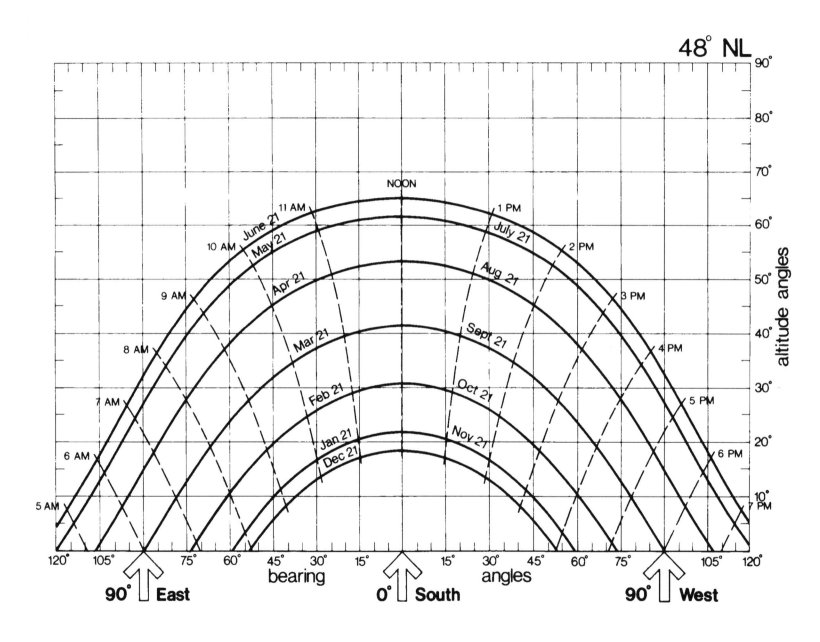

48° NL

altitude angles

90°
80°
70°
60°
50°
40°
30°
20°
10°

NOON

June 21
May 21
Apr 21
Mar 21
Feb 21
Jan 21
Dec 21

July 21
Aug 21
Sept 21
Oct 21
Nov 21

11 AM
10 AM
9 AM
8 AM
7 AM
6 AM
5 AM

1 PM
2 PM
3 PM
4 PM
5 PM
6 PM
7 PM

120° 105° 90° 75° 60° 45° 30° 15° 0° 15° 30° 45° 60° 75° 90° 105° 120°

bearing angles

90° East   0° South   90° West

B. maps and declination

Most people believe that a compass needle points to the north magnetic pole, but the Earth's magnetic field is the effect of complex convection currents in the magma. The compass actually points to the sum of the effects of these currents at your location. In other words, it aligns itself with the magnetic lines of force. Other factors, of local and solar origin, further complicate things. The overall result is that 'magnetic north' – where your compass points – is dictated by 'declination'.

The 0° declination (agonic) line passes west of Hudson's Bay, Lake Superior, Lake Michigan and Florida. North America stretches east and west from this imaginary line on the globe. East of the line, the needle points west of true north, and has a westerly or negative declination. West of the line, the needle points east of true north, and has an easterly or positive declination. The further north you go, and the closer you are to the north pole, the weaker the magnetic field and the more erratic a compass becomes.

Declination west, turn dial west (counterclockwise: add)
Declination east, turn dial east (clockwise: subtract)

Magnetic declination not only varies from place to place, but also with the passage of time, as the currents deep within the Earth change strength and direction. The magnetic declination of any location will change slowly over time, making older maps inaccurate. This is an important consideration when using magnetic bearings from old charts or metes (directions) in old deeds for determining solar south, or for locating places (or your site boundaries) with any precision.

Your map may state (or even illustrate) the local declination. Survey maps are primarily drawn on a 'magnetic north' grid, and show only one arrow, pointing to magnetic north. Most topographic maps include a small diagram with three arrows: magnetic north, true north and Universal Transverse Mercator grid north. Make sure to look at the date of any map you consult.

If your map does not show declination, you have two options: you can consult a general 'isogonic' chart for a rough estimate of the local declination (within a few degrees), or you can use a prediction model of magnetic declination for a given location developed by NASA. A map is sure to be months or years out of date, but the model is built around all the information available for a five-year period. The model will usually be more accurate than a map. National geological surveys or geographical institutes, local surveyors and airport control towers should provide the current declination, given your location in latitude and longitude, or just the name of a town.

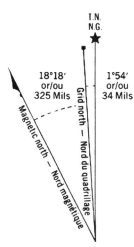

Above is an example of the type of symbol that shows magnetic, grid and true north on maps. Having this information, you can use a compass without a declination adjustment feature by adding or subtracting your declination from the compass reading.

Online resources:
Isogonic charts of Canada
geomag.nrcan.gc.ca/

Declination Calculator for Canada
geomag.nrcan.gc.ca/apps/mdcal_e.php

Current model of declination
www.ngdc.noaa.gov/geomag/magfield.shtml

C. insulation values

The insulative qualities of a material can be measured by its 'resistance' to heat flow, commonly known as the R-value of a material. In metric terms, the unit is RSI or 'U' value, which is the inverse of the R-value or RSI (make sure you know which scale is used or your calculations will be off).

The total insulating value of a wall, ceiling or floor assembly is the sum of the R-value or U-value of all components, including an insulating layer of air on the interior and exterior surfaces.

Adding the components of a building assembly together will give you a nominal insulation level, but will not account for differences in insulation levels between the framing and the cavity insulation. There are energy modelling software tools that can give you a better idea of the exact insulation levels by taking into account air sealing levels and the percentage of framing in the house. These have a learning curve associated with them, but aren't too difficult to master.

Following are some common building materials and their insulating values so you can determine nominal insulation levels for your project.

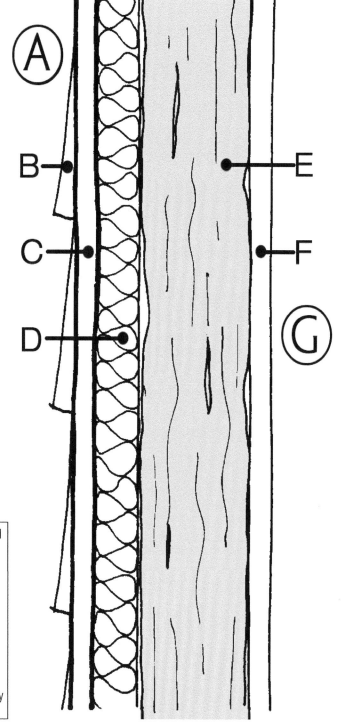

This cross section of an exterior wall shows:
A = outside air film
B = siding
C = sheathing
D = insulating sheathing
E = batt insulation
F = drywall
G = inside air film

The sum of all the insulating values indicates the total insulating capacity of the wall.

material	k-value (RSI)	R value
general		
outside air film	0.03	0.17
9mm (⅜in) sheathing	0.08	0.45
25mm (1in) solid wood	0.22	1.25
200mm (8in) concrete	0.26	1.49
12mm (½in) gypsum board	0.07	0.42
19mm (¾in) interior wood finish	0.17	0.94
walls		
inside air film	0.12	0.69
12mm (½in) min. air space	0.17	0.98
12mm (½in) air space/reflective surface	0.47	2.66
25mm (1in) glass/mineral fibre	0.52	2.97
25mm (1in) expanded polystyrene	0.69	3.96
25mm (1in) extruded polystyrene	0.87	4.96
25mm (1in) rigid glass fibre	0.74	4.25
12mm (½in) hardboard siding	0.13	0.73
19mm (¾in) drop siding	0.19	1.06
100mm (4in) clay brick	0.07	0.42
15mm (⅝in) stucco	0.02	0.12
metal/vinyl siding with backing	0.25	1.41
ceilings		
inside air film	0.11	0.60
100mm (4in) cellulose fibre	2.53	14.45
100mm (4in) wood shavings	1.69	9.66
100mm (4in) blown glass fibre	1.62	9.26
100mm (4in) vermiculite	1.44	8.23
asphalt shingles	0.08	0.45
wood shingles	0.16	0.94
floors		
inside air film	0.16	0.93
resilient floor coverings	0.01	0.08
carpet (rubber/foam underlay)	0.23	1.29
carpet (fibre underlay)	0.37	2.09
19mm (¾in) hardwood flooring	0.12	0.69

a primer on insulating values: R, k, U… what do they all mean?

R-values and k-values (also known as RSI, as in R-value, System Internationale) describe how good a conductor of heat the insulating material is.

The R-value or thermal conductivity is measured in square metre kelvins per watt ($m^2 \times K/W$), or $ft^2 x°F \times h/Btu$ in North America. The higher the number, the better the insulation. R-values are used in the US and Canada, although RSI is the unit referred to in Canada's metric system.

The k-value (RSI) is measured in Watts per metre per degree Kelvin (W/mK). The lower this number, the better the insulation.

The U-value is a measure of the overall ability of a building assembly to prevent heat loss, and is measured in watts per square metre Kelvin (W/m^2K). The U-value takes into account all the k-values of the various parts of the roof/wall structure, as well as the factors determining the transfer of heat from inside the room to the roof, and from the roof to the outside air.

Some building regulations and energy efficiency standards require building assemblies to achieve a U-value. This is because the U-value refers to heat transfer per square metre, while the R/RSI/k value is determined by the thickness of the material alone. So, because a pitched roof covering a house with a $100m^2$ ($1100ft^2$) footprint has more surface area than a flat roof covering a house of the same size, it will lose more heat than a flat roof. Thicker insulation is required for a pitched roof to achieve the same U-value.

Here are conversion factors:
To get RSI/mm, multiply R/inch by 0.00693
To get RSI, multiply R by 0.1761
U is 1/R, and R is 1/U

D. heat loss calculations

A heat loss calculation determines the size of the auxiliary heating system required for your house. The heating load will be highest on a winter night when the outside temperature is at its lowest and the sky is clear. The system you choose should have a capacity slightly larger than your heat loss calculation to compensate for really cold periods in the winter. Installing high-efficiency, properly sized heating equipment will save both fuel and money.

PROCEDURE

1. Calculate the total area of your ceilings, walls (minus openings for doors and windows), the above grade foundation wall, the below grade foundation wall, the foundation floor, the windows and the doors. If there are any other parts of your house that aren't listed here, but will affect the heat loss of your building, include these areas as well.

2. Using the table in Appendix C, calculate the total insulating value (R/RSI/K) of the assembly of each part of your house, including the inside and outside air films. You will need to allow for framing, which has a lower insulating value than an insulated wall cavity, and makes up about 15 per cent of your wall and ceiling area. Reduce your calculations accordingly.

3. Calculate the interior volume of your house (this is for your air leakage heat loss).

4. Look up the January design temperature for your location. (See sidebar at left for details on finding climatic information).

Contact your local weather office or government building standards branch to find the design temperature for your area. Design temperatures and degree days are also listed in the current National Building Code, which is available at public libraries in the reference section.
You can also find most of the data you need for your heat loss calculation at: eosweb.larc.nasa.gov/cgi-bin/sse/
If the link doesn't work, search for *NASA Surface meteorology and Solar Energy*
You will need to know the latitude and longitude of your site.

HEAT LOSS CALCULATIONS WORKSHEET

calculations

note: a normal inside temperature is 20°C

EG:
$(-16° - 20° = 36°C$ DIFFERENCE$)$

$$\text{equation \#1:} \quad \text{heat loss through building surface} = \frac{\text{area of surface} \times \text{difference between inside \& outside temperatures}}{\text{RSI (R) value of surface}}$$

ceiling heat loss
$$\frac{84 \times 36}{7.0} = 432W \quad \_\_\_ \times \_\_\_ = \_\_\_ W$$

wall heat loss
$$\frac{67 \times 36}{3.5} = 690W \quad \_\_\_ \times \_\_\_ = \_\_\_ W$$

above grade foundation wall heat loss
$$\frac{25 \times 36}{3.5} = 257W \quad \_\_\_ \times \_\_\_ = \_\_\_ W$$

below grade foundation wall heat loss
$$\frac{76 \times 25}{3.5} = 543W \quad \_\_\_ \times \_\_\_ = \_\_\_ W$$

floor heat loss
$$\frac{80 \times 15}{1.8} = 667W \quad \_\_\_ \times \_\_\_ = \_\_\_ W$$

window heat loss
$$\frac{12 \times 36}{0.5} = 864W \quad \_\_\_ \times \_\_\_ = \_\_\_ W$$

door heat loss
$$\frac{2 \times 36}{1.8} = 40W \quad \_\_\_ \times \_\_\_ = \_\_\_ W$$

BELOW GRADE WALLS & FOUNDATIONS HAVE SMALLER TEMPERATURE DIFFERENCES

$$\text{equation \#2:} \quad \text{heat loss through air leakage} = (0.36°) \times \text{volume of shelter} \times \text{air leakage rate in volume changes/hrt} \times \text{difference between inside \& outside temperatures}$$

air leakage heat loss
$$0.36 \times 437m^3 \times 0.2 \times 36 = 1132W \quad (0.36°) \times \_\_\_ \times \_\_\_ \times \_\_\_ = \_\_\_ W$$

total heat loss
(ADD ALL HEAT LOSSES AND AIR LEAKAGE)
4625W

• average heat gain
-500W   -500W

house total:
4125W

Use the following equations to calculate the heat loss for your design. Equation #1 gives you heat loss through the various building components. Equation #2 gives you heat loss through air leakage. This simplified calculation does not account for stored solar gain, so your system capacity might still be oversized.

Find the difference between your inside temperature (18° to 20°C/ 68° to 70°F) and the January design temperature (from step 4).

Equation #1: Heat loss through building component.

$$\frac{\text{SURFACE AREA } * \text{ TEMPERATURE DIFF.}}{\text{R-VALUE}}$$

Ceiling

(_____m² or ft² * ____°C or F) / RSI or R-____= ____W or Btu/hr

Wall heat loss

(_____m² or ft² * ____°C or F) / RSI or R-____= ____W or Btu/hr

Above grade foundation heat loss

(_____m² or ft² * ____°C or F) / RSI or R-____= ____W or Btu/hr

Below grade foundation heat loss

(_____m² or ft² * ____°C or F) / RSI or R-____= ____W or Btu/hr

Floor heat loss

(_____m² or ft² * ____°C or F) / RSI or R-____= ____W or Btu/hr

Window heat loss

(_____m² /ft² * ____°C/ F) / RSI/R-_____= _____W or Btu/hr

Door heat loss

(_____m² /ft² * ____°C/ F) / RSI/R-____ -= _____W or Btu/hr

Equation #2: Heat loss through air leakage

A			B	
ENERGY FACTOR	HOUSE VOLUME	*	AIR LEAKAGE RATE	TEMPERATURE DIFFERENCE

A: Equation #2 note: The energy factor represents the amount of energy required to raise one cubic unit (metre or foot) of air by 1 degree(C or F). Use the factor 0.36 for metric calculations, For imperial calculations, use the factor of 0.018.

B: Air leakage rate for standard construction is between 3 and 5 AC/hr; for R-2000 it is 1.5 and for the PassivHaus standard it is 0.6. Experiment with various rates to see how they affect your energy use. Always go for the lowest air change rate possible.

Total heat loss (add all heat losses and air leakage): ___W or Btu/hr

There are a number of heat sources within your house, including people, appliances and lights. Each person provides about 75 Watts of heating energy, appliances provide between 200 and 400 Watts. The average home has about 500 Watts of internal heat gains.

Internal heat gains: _____W or Btu/hr

Overall heat loss:
Subtract the internal heat gains from the total heat loss:
_____W or Btu/hr

A heat loss calculation is usually done by a heating contractor, but superinsulated and passive solar houses may result in drastic oversizing of a system.

Be sure you use either metric or imperial units throughout your calculations. If you are using metric, your energy units will be in Watts. If you are using imperial, your energy units will be in Btu/hr. To convert from one to the other (you may have to, depending on your energy source): 1watt = 3.414Btu.

183

E. sizing a PV system

ESTIMATING ANNUAL ELECTRICAL ENERGY LOAD
You need two pieces of information to determine your annual energy load: how much power each appliance, light and electronic device draws in watts, and how long, in hours, each of them runs on a daily (or weekly) basis. Most equipment will have a 'faceplate' on it, usually on the back (or the bottom) that shows the electrical specifications.

The sample worksheet (below) shows how this information can be laid out. Typical power ratings for appliances and equipment are listed in the table (centre column), but if you have the appliances at hand, you should use the faceplate specifications. Note any appliances that can be run on DC power (more of a challenge for grid-connected homes than off-grid homes, but still a viable way of reducing your energy load).

Typical Power Ratings – 115V AC loads		
appliance/equipment	power rating (watts)	annual kWh
refrigerator		100 to 500
freezer		100 to 500
dishwasher (energy use to run machine only)	1300	292
clothes dryer	4000	500
clothes washer (energy use to run machine only)	500	100
microwave	1000	
60 watt incandescent bulb	60	
24 watt compact fluorescent bulb	24	
portable fan	120	
coffee maker	900	100
toaster	1100	
hair dryer	1000	
iron	1000	
digital clock	2	
transistor radio	5	
radiophone (idle)	4	
radiophone (transmitting)	100	
portable phone	3	
answering machine	6	
14" colour TV	90	
VCR	30	
computer	15 to 200	16 to 200
printer	10 to 300	2 to 100
portable vacuum cleaner	800	

annual kWh includes automatic on/off cycling, multiply by 1000 to get watt hours, divide by 365 to get daily consumption.

appliance/equipment	power rating (watts)	annual kWh
power tools		
drill	300	
circular saw	400 to 4000	
furnace fan	350	1100
water pump	400	150
block heater	500	180
12V DC loads		
auto stereo	6	
digital clock	5	
25W incandescent bulb	25	
25W equivalent fluorescent	25	
colour TV, 2 hours/day	60	
toaster	1100	
15cm blade ventilation fan	24	
power tools		
drill	144	
circular saw	200 to 1000	
air compressor	60	
water pumps		
13L/minute automatic demand	90	70
11.6L/min	36	26
7.5L/min	18	13

PEAK POWER

Knowing your annual energy use takes you halfway through the process of determining the size of your system. Many appliances (including big draws like refrigerators and water pumps) cycle on and off. Small appliances and lighting are not in constant use. Water pumps and power tools have a start up 'surge' to get the motor going. To properly size your system you must estimate peak power consumption, even though it is unlikely that all of your equipment and appliances will be turned on at once.

From your list of appliances and equipment, add up the rated wattage of the most power-hungry appliances that could be operating at the same time.

CALCULATING BATTERY STORAGE CAPACITY

Batteries are rated in amp-hours. Take the estimate of your daily electrical load (which is in watt-hours). Assume three to five days will be the maximum amount of time without decent solar gain. You will require your daily load multiplied by that number of days. To find the number of amp-hours of storage you need, divide the watt-hours of your three to five day load by the voltage supplied by the batteries.

A battery can provide 6, 12 or 24 volts. Each battery has a specified depth of discharge (DOD): DOD is the amount of the total capacity of the battery that is safely available. Draining a battery beyond its DOD causes permanent damage. So, your amp-hours of storage need to be increased to compensate for the DOD.

Example:
1500 watt-hours daily load x 3 days = 4500 watt-hours
4500 watt-hours/24V battery = 187.5 amp-hours
compensate for a DOD of 50 per cent: 2 x 187.5 = 375 amp-hours of storage required (round up to 400 to be safe).

WHAT WATT?
The terms used in electrical systems can be a bit confusing. Here's what they represent:
Amp (Ampere): the rate (speed) of electrical flow. Wiring is rated according to how many amps it can carry.
Volt: the 'pressure' pushing the amps through the wire.
Watt: the combined speed and pressure creates the power available for use in your system.
Here's how they work together:
Watts = Volts X Amps.

appliance/equipment	AC	DC	rated wattage	hours per day	hours per year	annual Wh
4-24W compact fluorescents	√		4x24=96	5	1825	175,200
water pump	√		400	1	365	146,000
14" colour TV	√		90	2	730	65,700
high efficiency fridge		√	250	3	1,095	272,750
total annual energy consumption						660,650
total daily load						1,810
peak power			836			

F. rural site considerations

A simplified version of the Earth's water cycle is as follows (see illustration at right). Energy from the sun evaporates water from lakes, oceans and other bodies of surface water. Water vapour builds up in the atmosphere, forming clouds. When the clouds become saturated, rain (or snow) falls onto the land or back into surface water. It then drains into rivers and streams, or percolates through the surface and subsoil to replenish the groundwater supply. The water trapped under the Earth's surface eventually seeps into lakes and oceans. Thus, the cycle continues.

Topography – the lay of the land – determines your site's natural drainage patterns. These patterns can give you much information about your site, both above and below the surface. Knowing about your site's drainage patterns at the beginning of your project can save you from potentially expensive problems.

The water cycle affects any building site, but when you rely on a septic system and a well, you need to pay more attention to what happens on your site. Some of the physical variables to consider when determining the role of the water cycle in your site design include: topography and drainage patterns, soil conditions and geological formations.

Where your well and septic are located on your property ties into the water cycle through drainage patterns, soil permeability and percolation rates.

TOPOGRAPHY AND DRAINAGE

If the rain falling on your site is taken as the 'start' of the water cycle, you can readily observe what effect it has on the land. Rain will either drain off or percolate into the soil, depending on the topography and soil type. What do you need to know to 'read' drainage patterns? One simple fact: water travels from high ground to low, taking the path of least resistance. Does the water flow...

...into gullies or natural ditches?

Gullies and ditches provide visual clues for the primary drainage patterns on your site and also show you the 'contours' or slope(s) of your site. Changing the contours of your site to accommodate your house design will add to

the cost of construction, and will have an impact on your property and the surrounding properties as well. Better to change the design to suit the site.

...does it pool on the surface?

Drainage is obviously a problem where water pools on your site. Avoid depressed or flat areas, as they will tend to flood during the wet season. Pooling rainwater can indicate a high water table, or a clay-based soil. Organic soil drains well. Clay based usually puddles. A boggy or swampy area acts as a sponge of sorts. Filling or compacting these areas means that the water must flow somewhere else – possibly into your basement!

...is it absorbed quickly?

Areas that absorb rainwater quickly are usually good building sites, as the surface water easily percolates down to the subsoil. Other important factors to consider before deciding on a site for your house include slopes, the type of soil on your site and its bearing capacity. The type of soil and percolation rate will also affect where a disposal site can be located.

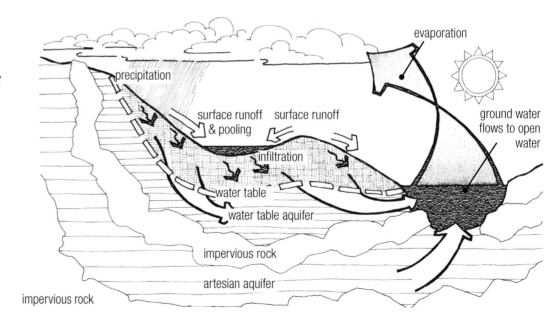

GROUNDWATER

The rain that doesn't drain off your site slowly percolates down through the soil, being purified of most pathogenic bacteria and other nasties as it goes. The purified rainwater, now known as groundwater, is stored in the underlying aquifer, a geological formation consisting of layers of sand, gravel, clay or rock.

There are two types of groundwater aquifers — water table aquifers and artesian aquifers.

Most residential wells tap into water table aquifers, basically an underground lake held in by a layer of impervious rock. These wells require a pump to bring the water to the surface. The level of the water table in most aquifers is seasonal and dependent on rain. A high water table can cause difficulties when excavating, as well as flooded basements or unstable foundations.

Artesian aquifers form below the layer of impervious rock that makes up the bottom of the water-table aquifer. The water contained in an artesian aquifer is confined under pressure between the bottom of the water-table aquifer and another, lower layer of impervious rock. When a well taps an artesian aquifer, the pressure pushes the water up the well. Some artesian wells have enough pressure to make a flow at the wellhead; others need to be pumped.

Water penetrates the soil at three levels:
held at surface by root systems of trees and other vegetation

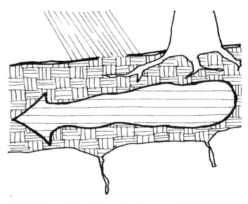

trapped by semi-impervious clay or fissured rock, moves laterally to surface drainage

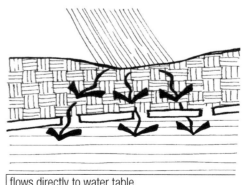

flows directly to water table

INDICATORS OF A HIGH WATER TABLE
Visual clues to a high water table include surface pooling, springs, discoloured or darkened soils, and the presence of moisture loving plants like willows, poplars, reeds and spongy mosses.

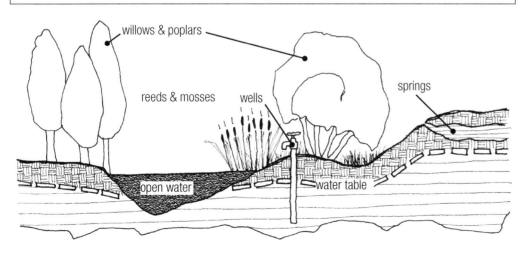

willows & poplars

reeds & mosses   wells   springs

open water   water table

A constantly flowing stream or creek is a primary drainage system. The banks of a stream or creek must remain in their natural condition in order to prevent erosion. Redirecting a stream or creek could wreak havoc on sensitive ecosystems further downstream. Because of this, you must have a permit from the appropriate government agency before you alter or fill in any watercourse.

All possible sources of contamination must be away and downhill (if possible) from your water source. A water source or in-ground storage downhill from your disposal field or sewer line may have a higher risk of contamination. Compost and animal-feeding areas should be downhill, all chemicals, toxic materials and petroleum products should be stored away from the water source, and vehicles (which can leak oil and gasoline into the ground) should be parked at a good distance.

A residential flow rate for wells of 32 litres (7 sp:gal) of water per minute is based on extremely high water use. If water-conserving measures are taken, a lower well flow can fulfill your water needs.

An inexpensive way to find out if the water table on your site allows you to dig a well instead of drilling is to have the excavator operator who is doing the foundation work on your house dig test pits in areas you suspect may prove to be productive. These test pits will only take 15 or 20 minutes, and can save you drilling costs.

WATER SUPPLY

On a site not serviced by a municipal water main, you must find a reliable source of potable (drinkable) water. This source can be found below or above the surface of your site: drilled or dug wells tap into groundwater sources, pumps tap into ponds or lakes, and cisterns collect and store rainwater.

The most common method of tapping groundwater is the drilled well.

A licensed, reputable driller who has worked in your area will know a lot about local groundwater conditions, the probable depth of your well, and the possible quantity and quality of water that you might expect.

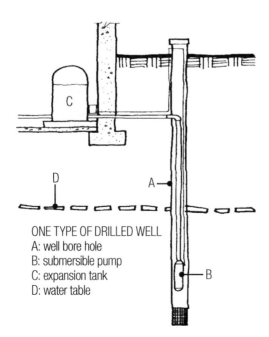

ONE TYPE OF DRILLED WELL
A: well bore hole
B: submersible pump
C: expansion tank
D: water table

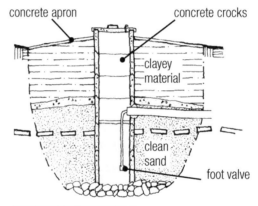

concrete apron   concrete crocks

clayey material

clean sand

foot valve

MACHINE DUG WELL

Machine or hand dug wells can be less expensive than drilled wells, but are sometimes not as reliable, as they usually only penetrate the upper water table aquifer. Seasonal fluctuations in the water level will affect the amount of water available to you to a greater extent than a drilled well. Dug wells are usually less than 15m (50ft) deep with a minimum diameter of 3.6m (3ft). A masonry or timber lining shores up the walls and protects the well from contamination.

If the production of your dug (or drilled) well is marginal, you can dig a swale (or french drain) to charge the water table uphill from the well. A swale is a long, shallow excavation that runs across a contour on a slope above the well. It catches surface runoff that is absorbed through infiltration into the soil.

HAND DUG WELL

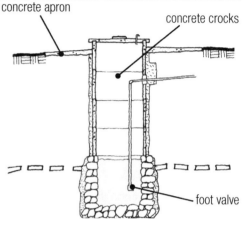

concrete apron

concrete crocks

foot valve

TAPPING AND PROTECTING A SPRING

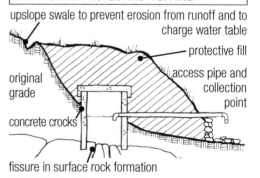

upslope swale to prevent erosion from runoff and to charge water table

original grade

protective fill

access pipe and collection point

concrete crocks

fissure in surface rock formation

Springs are either gravity-fed or artesian, with one or more surface points. To protect against contamination, install a large covered crock or liner, extend an access pipe from below the water level in the crock or liner to a collection point, and bury the whole structure under a mound of fill, leaving the collection point accessible. Even with this protection, contaminants can reach springwater from many different sources as it comes up from the aquifer, so periodic testing for bacteria levels is recommended.

CISTERNS

A constant supply of water depends on the reliability of the source. Where groundwater is in short supply, or is contaminated, rainwater can be used instead. Collecting rainwater for use in houses has a long history: huge cisterns have supplied water to the city of Istanbul since the fourth century. Before rural electrification and deep-well pumps, cisterns were a common water source for Canadians.

A cistern system can be very basic: gutters and downspouts collect the rainwater from the roof to the storage tank and a pump moves the water from the storage tank into the house through a standard plumbing system. Cisterns can be external to the house or designed into your foundation, so that the top of the tank becomes part of the floor of your house. Building or plumbing inspectors may require engineer-approved plans for a cistern. The size of your cistern depends on the amount of rainfall, the area of the collection surface (your roof) and the consumption patterns of your household.

Your local environment department or agricultural extension office will have annual rainfall figures for your area, and your local department of health or the environment will likely have some information about cisterns.

Rainwater collected directly from your roof will be soft, which is great for bathing and laundry uses, but soft water lacks essential minerals. This means that even though your cistern water meets health standards, it may not be recommended as a primary source of water for cooking and drinking without an adjustment to the pH level.

Also, to reduce the risk of contamination, you should include some method of disinfection, typically an ultraviolet light or chlorination. Any unwanted tastes in your cistern water can be removed by installing an activated carbon filter.

Ultraviolet disinfection units and/or activated carbon filters should be used on any water system.

A SIMPLE CISTERN SYSTEM

A typical example of a cistern, with a rudimentary 'roof washer' which acts as an inexpensive pre-filtering system by collecting the first flow of water off the roof, (contaminated with dirt, bird faeces, etc.). Once the drum is full, the rainwater flows into the cistern.

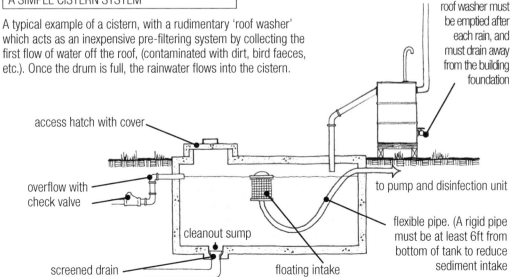

access hatch with cover

overflow with check valve

cleanout sump

screened drain

floating intake

roof washer must be emptied after each rain, and must drain away from the building foundation

to pump and disinfection unit

flexible pipe. (A rigid pipe must be at least 6ft from bottom of tank to reduce sediment intake

WASTE DISPOSAL

If your site is not serviced by a municipal sewer, you will have to dispose of your waste on site. Health regulations will, to a certain extent, govern where and how you can build on your site. With this in mind, the most accepted and economical system is a septic tank with a disposal field.

If your lot is not pre-approved, you must apply for a permit to install an on-site sewage disposal system. A qualified person then assesses your site and designs an appropriate system, based on lot size, soil conditions, water table levels, slope and potential loading.

SEPTIC TANK

Your permit will show the minimum size septic tank but you can increase the life of your system by installing an oversized single compartment tank. A double compartment tank will give you better waste treatment. When you buy your tank, make sure that it comes from a licensed manufacturer.

DISPOSAL FIELD

Topography affects the placement and layout of a septic field, but soil conditions determine the type and size necessary. The best location for your disposal field will be at or near the crest of a gentle slope. Areas below the crest may require

up-slope trenches to catch and divert surface run-off and/or groundwater from the field. Steep slopes combined with poor soil conditions make the design, installation and operation of an acceptable on-site disposal system difficult or even impossible.

The overall slope on your site will determine not only the type and design of your disposal field, but whether or not pumping is required to move the effluent from the house to the septic tank, and/or to the disposal field.

How well your disposal system works depends on the ability of the soil to absorb and filter the liquid sewage. Depth of soil and the relative percolation rate (how quickly water flows through saturated soil) are two of many variables considered in a disposal field evaluation.

Other considerations in determining the location of your disposal field include neighbouring wells and watercourses, which tie into the water table. Where sites are located one above another on a slope, this requires some long-range planning.

A leaching chamber disposal field is a system of manufactured concrete units designed to replace the distribution pipe and bed of a standard area or contour field.

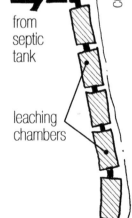

from septic tank

contour line

leaching chambers

TYPICAL LAYOUT OF A SEPTIC TANK AND DISPOSAL FIELD SYSTEM

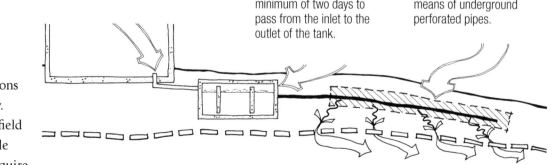

The building sewer is a water tight pipe that carries the raw sewage from the building to the septic tank, usually no less than 4in in diameter.

The septic tank is a settling tank where the raw wastewater is retained long enough for solids, fats and greases to settle to the bottom of the tank. For satisfactory performance, wastewater should take a minimum of two days to pass from the inlet to the outlet of the tank.

The soil absorption system (disposal field) carries the overflow of liquids from the septic tank and disperses them over a large area by means of underground perforated pipes.

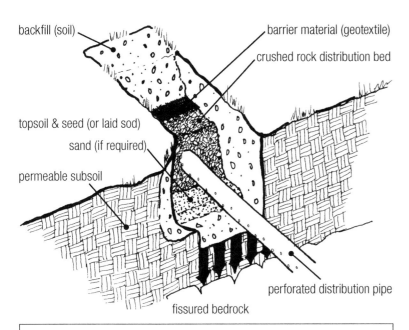

backfill (soil)

barrier material (geotextile)

crushed rock distribution bed

topsoil & seed (or laid sod)

sand (if required)

permeable subsoil

perforated distribution pipe

fissured bedrock

DISPOSAL FIELDS

There are three general types of disposal fields: area bed, multiple trench and contour trench. Area bed and multiple trench fields have been the typical layout for many years. Contour trench fields are relatively recent, and may not be used in your area.

INTERCEPTOR DRAINS

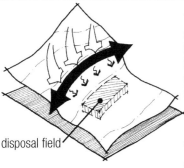

disposal field

These drains, or swales, run along contours upslope from septic fields. Most of the surface runoff and infiltration is carried away through a 100mm (4in) pipe that is covered with a layer of sand or gravel.

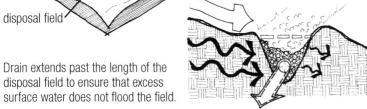

Drain extends past the length of the disposal field to ensure that excess surface water does not flood the field.

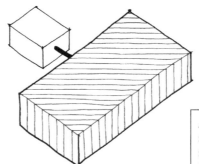

AREA FIELD: An area bed is a shallow rectangular excavation, suitable for deep soils of medium to high permeability on flat or gently sloping lots with a low water table. The long edge of the rectangle is oriented across the slope. This type of bed tends to become saturated beneath the centre of the field.

TRENCH FIELD: A multiple trench field is a rectangular net-work of shallow hand or machine dug trenches with the greatest dimension oriented across the slope. More efficient than an area bed, it is also suitable for deep soils of medium to high permeability on flat or gently sloping lots with a low water table.

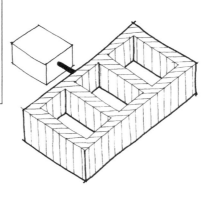

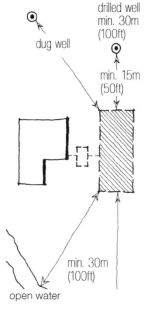

dug well

drilled well min. 30m (100ft)

min. 15m (50ft)

min. 30m (100ft)

open water

CONTOUR TRENCH: is a relatively narrow and shallow excavation dug along a contour (points of equal elevation). The effluent disperses into the subsoil and is filtered along the greatest possible length across the contour.

In most jurisdictions, the following clearances are applicable. Check with your local bylaws and provincial requirements to be sure.

The minimum clearance from a drilled well to a septic tank or a disposal field is 15m (50ft).

The minimum clearance from a dug well to a tank or field is 30m (100ft).

The minimum clearance from disposal field to a watercourse is 30m (100ft).

G. earthworks

You can use earthworks to reduce the energy required to heat and cool your house by berming north walls; to repair erosion damage and control drainage through contour banks, ditches, terraces and swales; to level out a portion of your site for placement of your house or road access, or to stop noise pollution with embankments.

The topsoil of newly constructed earthworks is extremely vulnerable: to stabilize it, plant fast-growing seeds or seedlings right after completion. This accomplishes two things: it prevents erosion, which can be severe on bare soils at only a 2 per cent slope; and it prevents weeds from colonizing the bare soil.

BANKS & BERMS

are the least expensive type of earthworks to construct, if the soil conditions are such that the bank can be made stable.

BENCHES

are flat cuts into hillsides along contours, usually for roadbeds. Where benches cross watercourses, they must have culverts or pipes installed or bridges built to allow the water to run under, instead of over, the bench.

Earthworks and retaining walls are a part of nearly every building project. When building a solar home, they become an important part of the energy conserving aspect of your design. The placement and construction of any earthwork or retaining wall should enhance the 'suntrapping' capacity around your house while also protecting the house from winter winds.

SLUMPING SLOPES

Every type of soil has an 'angle of repose'. As the sides of a bank or berm want to slump naturally, you can achieve better stabilization if you dish the sides as you build, making a concave profile inside the angle of repose.

Angles of repose

Recommended proportions of slopes for stable angles of repose over a range of soil materials.

Gravels
slope = 1:15 (37 per cent)

free-draining clay
slope = 1:2 (30 per cent)

sands
slope = 1:3 (21 per cent)

wet clays and silts
slope = 1:4 (16 per cent)

TERRACES

are most commonly seen in gardens. They must be built on stable soil, with good drainage so that water cannot be trapped and build up hydraulic pressure. Terraces are built from the lowest level up, on slopes no more than 30 per cent (10 to 18 per cent is ideal). The topsoil from each level is cleared onto the level below it, with that from the lowest level stockpiled and carted up to the top level.

DRAINS

run across the slope (along the contour line). The base of drains should be compacted to minimize the amount of water that soaks into the ground. Diversion drains are gently sloping drains used to lead water away from streams and valleys into storage and irrigation systems. Diversion drains are built to flow after rain, taking excess water away from one area to another. Interceptor drains take overland flows of water and direct them to streams or valley runoffs.

SWALES

are large, hollow or broad drains that allow surface water to pool and then soak into the ground. The base is loosened to allow water infiltration. Swales can be used to charge marginal wells or aid irrigation. Water enters the swale from roads, roof areas, tank overflows or diversion drains. The water is filtrated as it moves through the soil, removing bacteria. To keep the collected water from evaporating, moisture-loving trees should be planted on the south bank to shade the swale.

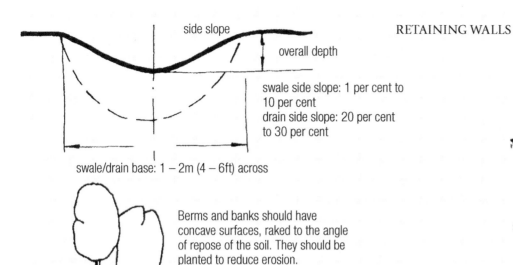

side slope

overall depth

swale side slope: 1 per cent to 10 per cent
drain side slope: 20 per cent to 30 per cent

swale/drain base: 1 – 2m (4 – 6ft) across

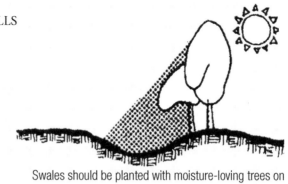

Swales should be planted with moisture-loving trees on the south slope

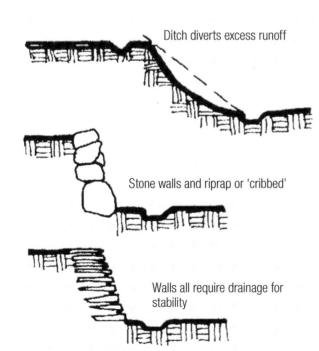

Berms and banks should have concave surfaces, raked to the angle of repose of the soil. They should be planted to reduce erosion.

Ditch diverts excess runoff

Stone walls and riprap or 'cribbed'

Walls all require drainage for stability

EARTHWORKS

Swales and drains are shallow excavations that are used to move water in different ways. The bottom of a swale should be 'ripped' gravelled or sanded to improve infiltration, while the bottom of a drain should be compacted to reduce infiltration.

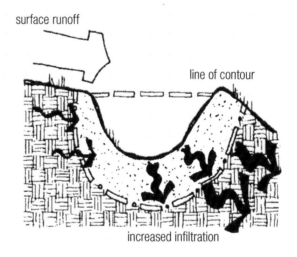

surface runoff

line of contour

increased infiltration

Machinery can be very destructive. Plan your earthworks so that the excavator or bulldozer operator has a specified route in and out of your site.

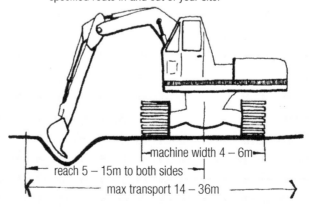

machine width 4 – 6m

reach 5 – 15m to both sides

max transport 14 – 36m

When carrying out any excavation, try to get the topsoil placed to one side so it doesn't mix with the sub-soil. This will save you buying topsoil to finish the job.

The lateral load on a retaining wall is generated by the pressure of the backfill against the vertical surface of the wall. This pressure is exerted on any type of retaining wall. It is the weight of the backfill beyond the angle of repose that causes the lateral load. Walls must also allow for drainage. Backfilling with clear stone or other free-draining material minimizes the build-up of water pressure behind the wall. If the water pressure is allowed to build up, the wall can become unstable and collapse.

CONCRETE WALLS

Concrete walls require steel reinforcing and weepholes for drainage. A reputable foundation contractor should be able to assist you in designing your wall safely and economically. An engineer or landscape architect must design large retaining walls. Concrete or concrete blocks can be faced with rock or brick.

RETAINING WALLS

Retaining walls are used to stabilize and modify natural and constructed slopes. They can be made of concrete, with several different footing types designed to support the lateral pressures exerted by the earth behind them. Building structurally sound concrete retaining walls requires some expertise.

Retaining walls can also be handbuilt with rocks, timbers or old tires. Although these types of retaining walls are easier to build than concrete walls, because of the nature of the materials as well as the limited anchorage they afford, they must be carefully designed and properly built. Handbuilt retaining walls under 1250mm (4ft) high are easy to build and require little or no reinforcement, but if you are planning a wall over 1250mm (4ft) high, you should enlist the services of an engineer or a landscape architect.

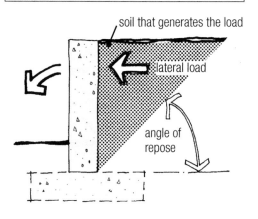

soil that generates the load
lateral load
angle of repose

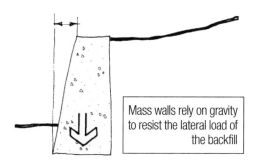

Mass walls rely on gravity to resist the lateral load of the backfill

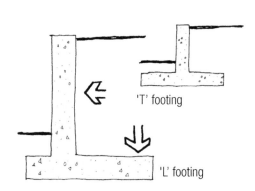

'T' footing

'L' footing

Cantilevered walls use the weight of the soil on the footing to resist the lateral load.

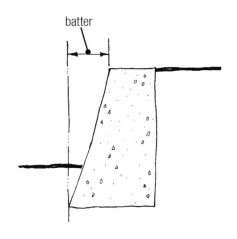

batter

'Batter' refers to the slight slope of the outward face of any type of mass wall.

Walls made with rocks about 80 x 150 x 360mm (3 x 6 x 14in) can be laid with or without mortar. 'Dry laid' walls must have a 'batter' (a slight slope) of 80mm per metre (1 inch per foot). Dry laid walls must have backfill put in as they are being constructed. This is a reasonably skilled job. If you decide to build a dry-laid wall, research the method and materials available and try some small practice projects before tackling a big wall. A mortared wall requires less skill than a dry-laid wall and can be built with different kinds of rocks. If the mortar is raked very deep into the structure, it will have the appearance of a dry-laid wall.

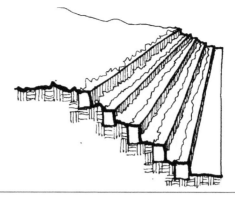

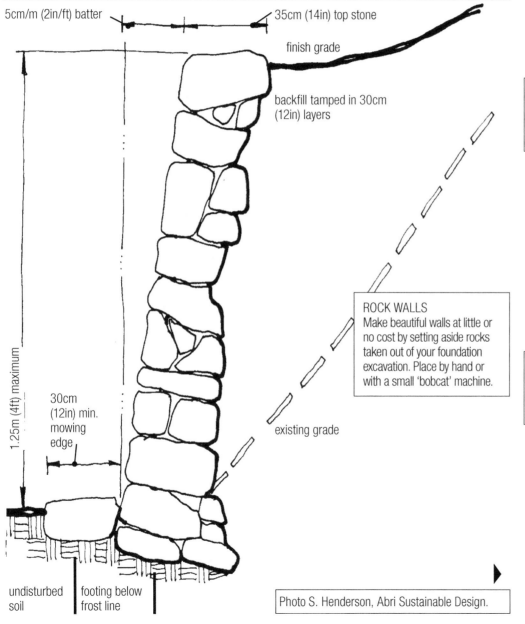

5cm/m (2in/ft) batter

35cm (14in) top stone

finish grade

backfill tamped in 30cm (12in) layers

1.25m (4ft) maximum

30cm (12in) min. mowing edge

existing grade

undisturbed soil

footing below frost line

ROCK WALLS
Make beautiful walls at little or no cost by setting aside rocks taken out of your foundation excavation. Place by hand or with a small 'bobcat' machine.

CRIBBING
Use railway ties, stones or old tires raked back at the angle of repose for the soil. Should be planted to avoid erosion.

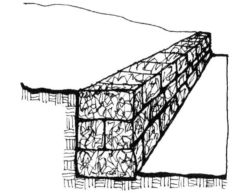

GABIONS
Fill sturdy wire baskets with stones and rubble to stabilize eroding gullies. Gabions resist flooding well.

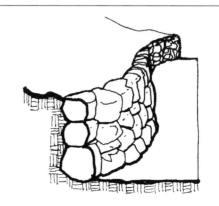

Photo S. Henderson, Abri Sustainable Design.

Always backfill hand built walls as you go; place the largest rocks or timbers at the base of the wall.

H. resources

Solar Nova Scotia (Solar NS)
www.solarns.ca
Publishers of the Canadian Solar Home Design Manual.
Provincial chapter of SESCI.

Canada Housing and Mortgage Corporation (CMHC)
www.cmhc.ca
Up-to-date information on building science and housing issues.

RETScreen International
www.retscreen.net
The RETScreen Clean Energy Project Analysis Software is a
unique decision support tool. The free software (available in multiple languages) can be used worldwide to evaluate the energy
production and savings, costs, emission reductions, financial
viability and risk for various types of Renewable-energy and
Energy-efficient Technologies (RETs).

**Natural Resources Canada (NRCan), Sustainable Buildings
and Communities**
www.sbc.nrcan.gc.ca

NASA Surface meteorology and Solar Energy
www.eosweb.larc.nasa.gov/cgi-bin/sse/
Weather and climatic data, all information on sunlight hours
etc. for sizing systems and heat loss calculations. You will need
to know the latitude and longitude of your site.

There is a constantly growing
body of online resources.
Use the following terms for
keyword searches: *climatic
housing, solar, passive
solar, solar thermal,
grid-connect, off grid, pv,
low impact housing , low
energy housing, carbon
neutral housing, deep
energy retrofits, passive
housing, EQuilibrium housing.*

US resources

American Solar Energy Society (ASES)
www.ases.org

Energy Efficient Builders Association (EEBA)
www.eeba.org

Affordable Comfort Institute (ACI)
www.aci.org

North Eastern Sustainable Energy Association (NESEA)
www.nesea.org

Rocky Mountain Institute
www.rmi.org

National Renewable Energy Laboratory (NREL)
www.nrel.gov
The facility of the Department of Energy (DOE) for renewable
energy and energy efficiency research, development and deployment.

Energy Efficiency and Renewable Energy (EERE)
www.eere.energy.gov
Department of Energy (DOE) website for information on energy
efficiency and renewable energy technologies.

Solar Rating and Certification Corporation (SCRCC)
www.solar-rating.org
Independent certification of solar water and swimming pool
heating collectors and systems.

EU resources

Carbon Trust
www.carbontrust.co.uk
An independent not-for-profit company set up by the UK government
to take the lead on low carbon technology.

ManagEnergy
www.managenergy.net/energyagencies.html
Website of the European Commission Energy, includes information
about nearly 400 energy agencies throughout Europe.

International Solar Energy Society (ISES)
www.ises.org
Comprehensive technical body dealing with the technology and implementation of renewable energy. Follow links to regional ISES chapters.

International Energy Agency Solar Heating and Cooling Programme
www.iea-shc.org
A collaborative solar heating and cooling research programme managed by solar experts from the US and the EU.

technical manuals (search online for best buys)

Builder's Guide to Cold Climates Details for Design
Joseph Lstibrek
A manual of design details by a leading building science expert and an expat Canadian. Also visit his corporate website at www.buildingscience.com/bsc/designsthatwork/cold/bestpractices.htm

Green From the Ground Up – A Builder's Guide
David Johnson and Scott Gibson
A good and up-to-date construction oriented book on building energy efficient and green homes. Fairly nuts and bolts coverage of what's green and energy efficient in foundations, framing, HVAC, insulation, electrical, plumbing, etc.

New Autonomous House
Robert Vale

The First Passive Solar Home Awards
US Department of Energy

The Passive Solar Construction Handbook
Steven Winter Associates, Inc
A book covering many construction details used in passive solar home designs.

The Renewable Energy Handbook
William Kemp (Canadian author, Canadian content)

The Solar House: Passive Heating and Cooling
Dan Chiras
Also visit his corporate website at www.danchiras.com.

The Superinsulated Home Book
J. D. Nisson and Gautam Dutt
This very good 1985 book provides a great deal of actual construction detail for superinsulated homes. Some information is outdated, but most is still relevant.

The Whole Building Handbook: How to Design Healthy, Efficient and Sustainable Buildings
Varis Bokalders and Maria Block

Solar Architecture in Cool Climates
Colin Porteous

Sustainable Solar Housing
(2 volumes) edited by S. Robert Hastings and Maria Wall

relevant magazines (online and in print)

Fine Homebuilding

Home Power Magazine

Journal of Light Construction

The Natural House

INDEX

T - #0913 - 101024 - C208 - 216/280/9 - PB - 9781844079698 - Gloss Lamination